Biodiversity in Agriculture

Kodoth Prabhakaran Nair

Biodiversity in Agriculture

Sustainability of Soil, Soil Fauna and Soil Flora

Springer

Kodoth Prabhakaran Nair
International Agricultural Scientist
Malaparamba, Kerala, India

ISBN 978-3-031-44251-3 ISBN 978-3-031-44252-0 (eBook)
https://doi.org/10.1007/978-3-031-44252-0

This Springer imprint is published by the registered company Springer Nature Switzerland AG
The registered company address is: Gewerbestrasse 11, 6330 Cham, Switzerland

Paper in this product is recyclable.

I dedicate this book to Pankajam, my wife, a Nematologist trained in Europe, but, one who gave up her profession, and, instead, chose to be a home maker, more than four decades ago, when we had our son, Kannan, who is now a doctor, and Sreedevi, our daughter, who is now an engineer. Pankajam is my all, and, she sustains me in this difficult journey, that life is. It is also dedicated to the memory of my late parents, my father, Kuniyeri Pookkalam Kannan Nair, an illustrious police officer, who served the British Police, and who was decorated with the King George V medal for bravery and honesty, and my mother, Kodoth Padinhareveetil Narayani Amma, daughter of the aristocratic Kodoth family of North Malabar, Kerala State, India, both of whom left me an orphan at a very young age, but, whose boundless love and blessings made me what I am today.
Sasha, our daughter's cocker spaniel, is a constant source of joy to us whenever we visit our daughter. This book is dedicated to her, as well.

Father Mother

India's great President, Late Dr. A. P. J. Abdul Kalam, launching the book Issues in National and International Agriculture authored by Prof. Kodoth Prabhakaran Nair, in Raj Bhavan, (Office of the Governor of the State), in Chennai, Tamil Nadu, India

Preface

The observance of the International Biodiversity day on May 22, 2023, yet, again, reminds mankind of the crucial role biodiversity plays in the world, be it in resolving the climate crisis, which along with the steep decline in biodiversity, all over the world, is, now, an existential threat to the future of mankind. Biodiversity, the rich variety of life forms, be it a plant, an animal, a microorganism, even the soil, and their interconnections with each other, and the environment, at large, is everywhere: inside our bodies as ubiquitous microbiomes, in our backyards, villages, towns, and cities, and in remote wild places, as well-organized communities or ecosystems. Maintaining and enhancing biodiversity on land and even the oceans is, perhaps, the least expensive mechanism to sequester not only carbon dioxide, but also other greenhouse gases, such as nitrous oxide, an offshoot of the highly chemical-centric soil extractive faming, euphemistically called the green revolution, from the atmosphere, which will cool Planet Earth and our oceans. In terms of its potential to heat the atmosphere, nitrous oxide is 300 times more potent than carbon dioxide, and it remains in the stratosphere for over 100 years (Nair, 2019c).

Mitigation of climate is, but, one of the several benefits of biodiversity. Biodiversity also fulfills our basic needs of life, like food, shelter, medicines, mental health, recreation, and spiritual enrichment. To mitigate the continuing decline in the quality of our environment, we will need to rely more and more on solutions which draw upon biodiversity, which can be termed as "nature-based solutions" to secure our future. It is biodiversity that will restore our vast sources of degraded lands in South Asia—a very sand environmental hazard caused by the green revolution. Biodiversity will restore our polluted rivers and oceans and sustain our agriculture, in the face of climate change. It is biodiversity that will form the basis of a new, and sustainable, "green economy". And, it is biodiversity that will inspire the future generations to opt for a more humane, equitable, just, and hopeful future, which accords primacy to the living world. In short, it will give primacy to the ideas propounded in the book "The Living Soil" by this author (2023a), where he places soil at the center of all life, plant, animal, and man, in fact conforming to the phrase he coined for soil, which is "**Soul of Infinite Life**", during his address in the delegates of the World Soil Science Congress in Hamburg, Germany, in 1986.

Despite the crucial importance of biodiversity which ultimately sustains all human endeavors, we have been poor stewards for caring and nurturing life on Planet Earth. Globally, we have failed adequately to conserve and efficiently manage our precious, irreplaceable natural heritage. Worldwide, biodiversity is declining, and, our last remaining, largely isolated ecosystems are degrading due to rapid changes taking place around them, such as loss of species, both plant and animal, climate stressors, and relentless human pressure, what this author has termed "**Anthropogenic Greed**". In India, for example, the Forest (Conservation) Amendment Bill, recently enacted in the Indian Parliament, will further weaken our resolve to sustain the remaining biodiversity.

How do we nurture and manage diversity is a central question we all must address to. In many ways, biodiversity is us and we are biodiversity. Hence, civil society must play a very critical role in sustaining biodiversity. A paradigm shift in the care of biodiversity is very long overdue; we must begin now, flowing from the above-mentioned International Biodiversity Day.

Let us first change the manner in which we manage our biodiversity. I shall take here the example from the vast Indian subcontinent. Currently, the main custodian of the natural world in India is the Indian Forest Service. But, the term "forest" used to describe the immense and unique natural heritage of India is, indeed, flawed. India's biodiversity is not only on land, but also in the water bodies, rivers, rivulets, deltas, and oceans. A rich array of India's ecosystems is in the form of grasslands, savannahs, alpine pastures, deserts, and other types of ecological communities. Even in the twentieth century, people had started talking about living organisms and the interconnectedness manifested as ecosystems and ecosystem services in multifunctional landscapes dominated by humans. In the twenty-first century, the basic terms, "forests" and "wildlife", have, but, limited meaning and usefulness.

We must think of multifunctional landscapes, where aspirations, beliefs, traditional knowledge, and direct participation of local communities are central to the notion of conserving and sustaining life on Planet Earth. Interestingly, in 2006, the Indian government enacted a law called "Forest Rights Act", which called for an increase in the stake of local/indigenous groups on ownership and management of biodiversity. Sadly, this Act simply remains on paper, even now, with no follow-up action. Seventeen years is a long wait, indeed! If biodiversity is everywhere, as it is, we must mainstream it into our daily actions in every development program, in every government department, and in every public and private institution.

How do we mainstream biodiversity?

The central goal of the proposed National Mission on Biodiversity and Human Wellbeing is to mainstream biodiversity, an idea proposed by Prof. Kamal Bawa, President Emeritus of the Bengaluru (India)-based Ashoka Trust for Research in Ecology and the Environment (ATREE), and Convenor of the Biodiversity Collaborative. India's leading conservation biologists, working under the umbrella of the Biodiversity Collaborative, based in Bengaluru (India), conceptualized the idea and developed a road map for the National Mission, named above, approved, in principle, by Prime Ministers (of India), Science, Technology and Innovation Council.

The Mission will enable India to meet critical challenges in climate change, natural and regenerative agriculture (Nair, 2023), ecosystem and public health using biodiversity and ecosystem services—usually referred to as "nature-based solutions". The ultimate goal of the Mission is to enhance and conserve India's biodiversity to foster well-being of Indians, in particular, more specifically to meet the United Nations Sustainable Development Goals (SNGs), related to poverty alleviation, nutrition, health, environmental protection, and, in the process, support a new era of "green economy", for Indians, in particular, and the world, at large.

People will be at the center of the Mission, the goal of which is to have all Indians engaged in the conservation and sustainable use of biodiversity, embed consideration of biodiversity in every development-oriented program of the public and private sectors, and arouse curiosity about nature and a sense of responsibility to safeguard biodiversity—and the world's very future—in the minds of every child and every human being. Undertaking such a pledge would be the most fitting celebration of mankind's precious and irreplaceable natural world.

This book is all about biodiversity—plant, animal, mankind, and soil—and its central role in sustaining life of all on Planet Earth.

Malaparamba, India Kodoth Prabhakaran Nair

Acknowledgements I greatly appreciate the efficient support of Ms. Margaret Deignan, Senior Editor, Springer. Mr. Karthikeyan Krishnan, Project Coordinator (Books), and his team, who did a marvelous job in bringing out this remarkable book. I am delighted to add a word of appreciation for Mr. Madanagopal Deenadayalan, for his conscientious and diligent work on proof correction of this important book.

References

Nair, K. P. P. (2019c). Combating global warming: The role of crop wild relatives for food security. Springer Nature, Switzerland, AG 2019.

Nair, K. P. P. (2023a). The living soil: A lifetime journey in understanding it for human sustenance. SpringerBriefs Environmental Science.

Contents

Chapter 1
Introduction

Evolution proves that life on planet Earth has moved forward by enriching itself by adding new forms. This is true of plant and human beings. When we keenly observe the biosphere, we notice that a variety of organisms, be it a plant, or an animal, or even man, each enriches itself during the passage of time. This very much applies even to the highly evolved human beings, starting from the smallest microorganism. There are five kingdoms, namely, Protista, Monera, Fungi, Plantae and Animalia. These five kingdoms encompass living organisms based on their cell type, be it, prokaryotic or eukaryotic, unicellular or multi-cellular, genetic recombinants through binary fission, asexual or sexual, or feeding mode of nourishment, be it autotrophic or heterotrophic. There is further division of the kingdoms into phyla (among animals), classes, orders, families, genus and species. In each of the kingdoms, there is huge diversity. Every species in every kingdom is unique in itself. It is this uniqueness which causes the biodiversity. These organisms in their multiple numbers, which can be called populations, organize themselves in specific communities. Each community, in itself, is unique, inasmuch as the variety of the species and operational environmental factors, and, inasmuch much as its structure and functions are concerned. There are innumerable genotypes in every community, with specific genetic characteristic, for all species of plants, animals, and microorganisms. On Planet Earth, biodiversity may be termed as the "greatest wonder" and very much the ecological integrity and sustainability of life on this planet.

How Can One Define Biodiversity?

It was Thomas Lovejoy who in 1980 coined the term "biological diversity" while writing in the monumental work Conservation Biology (Soule and Wilcox, 1980). It was W.G. Rosen (Wilson, 1988) who coined the word "Biodiversity". The field of biodiversity, with interdisciplinary disciplines, enlarged rapidly, in just about three decades (Nautiyal et al., 2012). The term "biodiversity" or "biological diversity"

K. P. Nair, *Biodiversity in Agriculture*,
https://doi.org/10.1007/978-3-031-44252-0_1

represents the variability among all living organisms, whether it is of plant origin, or animal origin, for all terrestrial and aquatic ecosystems. It includes, varieties of all organisms, as well as, the complexity the create. There is enormous biodiversity on planet Earth. Biodiversity, in essence, is the sum total of all genes, all species, and all ecosystems of the biosphere, depicting the beauty of planet Earth.

If one considers the distribution of the species, it can be observed that not all species occur in the same place. The environmental factors which influence a site, with regard to ambient temperature, humidity, soil characteristics etc., all of which determine the prevalence of a specific species in a locality/site. This is the reason why biodiversity varies from region to region, area to area, and, place to place in the same area/region. Hence, taking into consideration all the habitats of plants, animals, and microorganisms, one can imagine the extent of the enormous amount of biodiversity on the biosphere of planet Earth.

In fact, the human species depends, exclusively, on nature's biodiversity for survival. Hence, human species, in itself, is an indistinguishable part of the nature's biodiversity. Biodiversity, is, in fact, a precondition for the survival and materialistic progress of the human species. Also, there is dazzling diversity among the human species, at the genetic level. Cultural variability in human species is also a manifestation of the biodiversity of nature.

When natural habitats are destroyed, a consequence of human greed, what *shall term "**Anthropogenic Greed**"* (human created greed), emanating from a spurt in human population, has become a major cause of biodiversity destruction and depletion, which, ultimately, when taken to uncontrollable heights, will result finally in the extinction of human species, along with it, all the other species.

What Are the Levels of Biodiversity?

The characteristic feature of life, be it human species or animal species, plants, or plant species, or even soils, there exists an enormous amount of genetic variation. When it comes to even soils, the same principle operates, as has been demonstrated recently by Nair (2023a, 2023b), in his most recent book titled "The Living Soil". The genetically distinctive individuals of every species live in a variety of ecological systems amidst complex ecological relationships. Hence, an understanding of biodiversity is crucial for the recognition of its multiple values and, consequently, its conservation. Under the "umbrella" of the biosphere, biodiversity prevails at the following three hierarchical levels:

1. Genetic Diversity
2. Species Diversity
3. Community and Ecosystem Diversity

With the above mentioned three distinctive characteristics, biodiversity prevails amidst the state of ecological balance and ecological integrity. The following discussion pertains to the three categories named above.

1. Genetic Diversity

There is a vast amount of genetic information in every individual species. For instance, *Homo sapiens* (the human species) has as many as 35,000–45,000 genes to display (though 20,500 genes according to one estimate). Among the plant species, rice (*Oryza sativa*), for instance, has 32,000–50,000 genes. The commonly researched Fruit fly (*Drosophila melanogaster*) has 13,000 genes, and, the human colon bacteria (*Escheria coli*, commonly referred to as *E. coli*, which causes many stomach disorders in human beings, such as, urinary tract infection, commonly known as UTI) has 4000 genes. These variability in gene number causes genetic variation among all species, be it humankind, animal kind or plant kind. However, though soils have not been found to contain any "genes", they, have algorithmic characteristics which could be explained mathematically, thus, giving soils the name "Living" (Nair, 2023a, 2023b). The varying genetic information determines certain specific characteristics of a species.

"Genetic Diversity" is the term that explains genetic (gene-specific) variation within a species. In humankind and animal kind, the genetic variation relates to chromosomal structures, in entire genes, that is, traits which determine specific characteristics, or, in alleles, that is, different variants of the same genes. Genetic diversity is vital for the population of a species to adapt to varying environmental conditions and respond to natural selection. The higher the degree of genetic diversity, the better the ability of the species to adapt to changed environmental conditions. A narrow base of genetic diversity results in uniformity. Genetically similar crops result in monocultures. A trend of raising crop monocultures is encountered in modern farming methods. The most illustrious example is the rice-wheat crop rotation of Punjab State, in India, as part of the so-called "green revolution", which has led to enormous environmental problems, in particular, soil related—degradation of soil resources. This aspect is separately discussed in the book "Intelligent Soil Management For Sustainable Agriculture—The Nutrient Buffer Power Concept" (Nair, 2019a, 2019b). As has been conclusively proved during the highly chemical centric farming, euphemistically called the "green revolution", though monocultures have provided higher levels of food grains for a short span of time, they are ecologically very vulnerable and are liable to be destroyed due to the onslaught of newer pests (both bacterial and fungal, also insect pests), which has been clearly explained by Nair (2023a, 2023b), in his most recent book "Extractive Farming or Bio Farming—which is a better choice for 21st century?). A species with a broader genetic base, that is, with a high degree of genetic diversity, has a high degree of resilience and, is, far more ecologically stable. Genetic variability within a species, or intra-species diversity, often registers an enhancement of environmental variability. Speciation, the evolution of new species, depends on the quantum of genetic variability. Speciation is vital to maintain a high degree of diversity at the level of species or communities. The higher the number of species in a community, the higher the level of genetic diversity it harbors and vice versa.

2. Species Diversity

When a specific region in the world has a variety of species, it is known as species diversity. All of the species of plants, animals and microorganisms, across all of the five kingdoms in the biosphere count as species diversity. The species are the most distinct units of biodiversity and have a specific role to play in the ecosystem. When a species is lost, it manifests greatly on the structure and functioning of an ecosystem in which it prevails. The species diversity of a region, or a specific geographical area, is determined based on the following two criteria:

1. The richness of the species in consideration
2. The equitability or evenness

The first point denotes the number of species existing per unit area. Following an increase in the size/area of the region/site, the species number also increases. On the whole, following the increase in species richness, the species diversity also increases. Higher the species richness, the greater the species diversity.

Equitability or evenness denotes the evenness in the number of individuals of a particular species. Where there is a higher species diversity, a higher evenness can be expected in a specific region/geographical site. The species evenness can be understood considering the following examples:

1. Sample area A: Three species of a microorganism of which, two are represented by one individual each, the third one has four individuals
2. Sample area B: Three same species of the above mentioned microorganisms each represented by two individuals
3. Sample area C: Three species of the same microorganism, a butterfly and a bird

From the above samples, it can be inferred that sample area B shows greater evenness than sample area A, with equal chances for the representative species to occur, which reveals greater/higher diversity in sample area B as compared to sample area A. Sample area C has taxonomically unrelated species, hence, they harbor the highest species diversity among all the three sample areas, A, B and C. Each sample area depicts an equal number of species, but, different number of individuals per species. Hence, the number of species, also, several individuals per species, determine the measure/extent of species diversity in a specific region or specific geographical site.

The term endemism refers to the ecological state of a species, being unique to a defined geographical location. These organisms which are indigenous to a region/geographical unit/site, would not be counted as endemic, when found elsewhere. In biodiversity, we denote that cosmopolitan distribution occurs which is an extreme case of endemism.

3. Community and Ecosystem Diversity

While species organize themselves as populations, populations, in turn, organize themselves as communities. All biotic communities are not similar. One would find enormous variations among them. Community diversity reflects variations in the biotic communities in which the species thrive. With regard to community diversity, three distinct entities, as follows are recognized.

A. Alpha Diversity
B. Beta Diversity
C. Gamma Diversity

Alpha Diversity: This diversity refers to the diversity within a community, that is, diversity of the organisms which share a common community. Within a community, both traits, richness and evenness or equitability are applied to measure the degree of diversity within a community.

Beta Diversity: Between communities, when diversity occurs, this parameter is referred to. Communities are never similar to each other. They change, along with the species. Not two or more communities contain the same species. The composition of species changes owing to certain environmental parameters, such as altitude, ambient temperature, moisture content etc. The more elevated the heterogeneity within a community, within a region or a specific geographical area/site, the greater the measure of beta diversity between them.

Gamma Diversity: This term refers to the diversity of the habitat or communities over a specific geographical area. This reflects the turnover rates of the species concerning the geographical distance between the sites of similar habitats or with enlarging geographical area. Gamma diversity is applicable only to large geographical areas. A high level of diversity is the basis of a higher productivity and stability of a community. A higher diverse community is comparatively healthier, vibrant, and, in the long run, more sustainable.

In an ecosystem, there is a community, which, indeed, is its biotic content. Both biotic and abiotic components constitute an ecosystem and they are reciprocal in interaction, with each other. Hence, ecosystem diversity is distinguishable from community diversity, inasmuch as the terms of the variations in the structure and function of an ecosystem are concerned. A variety of niches, such as, trophic levels and ecological processes, sustaining the food chains, food webs, and nutrient cycles are described by an ecosystem diversity.

When an ecosystem is complex, it provides a large number of ecological niches which provide an opportunity for the organism living within the community the possibility for many activities. Larger the niches, greater the ecosystem diversity. When environmental factors lead to isolation of an geographical area, it would only harbor limited number of species. For instance, if an island loses its species, say due to an extreme weather aberration like the tsunami, it would be extremely difficult for it to regain that species located in another area. A dominant species in an ecosystem dominates natural resources at the expense of others, which will impose restrictions on an enhancement of the biodiversity of an ecosystem. Ecotones or transition areas show a relatively higher degree of biodiversity, which is called an "edge effect". "Keystone" species determine the ability of a large number of other species to persist in the community, which plays a crucial role in ameliorating the environment for many other species which enriches the ecosystem diversity. It is interesting to note that all top carnivores, such as, a crocodile, a tiger, and a lion, are known as keystone species, which, in truth, are regulating the population density of other animal species, in a given ecosystem. A lion or a tiger will prey on deers to keep their populations under check, from vastly multiplying. If a keystone species vanishes, from the ecosystem, it would trigger the process of ecological degradation.

It is also interesting to note that geological history takes place to be a decisive global factor enhancing or restricting/curbing biodiversity in a region. It is this factor, that, for instance, arctic environments did not permit most of the species, both animal and plant, including human, to flourish, but, on the other hand, well established ecosystems, as for instance, rainforests, allowed a high degree of biodiversity to flourish in their environments. Biotic and abiotic factors create habitat stresses creating conditions unfavorable for the biodiversity to flourish in a specified area. Additionally, there are some other interesting terms concerning community and ecosystem diversity. These are the following.

1. Point Diversity: This term refers to the diversity, at the smallest scale, as for instance, in a microhabitat
2. Epsilon Diversity: This, on the other hand, refers to the diversity of a group areas of gamma diversity, which is also known as regional diversity

What Are the Gradients of Biodiversity?

Across the Earth, there is no uniformity in biodiversity patterns. Apart from some local natural anthropogenic factors, two factors which markedly affect biodiversity, are altitude and latitude. Variations in natural biodiversity occur following the changes in altitude and latitude. Biodiversity gradually increases from high to low altitude. This means, there is a gradual increase in biodiversity from the poles to the equator of planet Earth.

The Effect of Latitude on Biodiversity

The climate in temperate regions is more severe and quite inhospitable for life. This is especially true of glaciated poles of the Earth. Species that evolved in this region tended to migrate toward favorable environments. Further, the annual period for growth of the plants is short. The photosynthetic base for the support of consumer species is quite narrow. Pressure of pests, and disease-causing organisms is not up to mark in cold climates. Hence, only a few dominant species prevail in temperate environments. One encounters the lowest levels of biodiversity in the cryosphere, (part of the planet earth having frozen water) where only a few animal species survive.

In the tropics, biodiversity is a lot more at lower altitudes, where ecosystems are more complex, heterogenous and stable, which is attributable to the following reasons:

1. Tropical areas are not glaciated, with a stable climate, which is hospitable to many species.
2. During evolution, tropical regions have encountered many cycles of temperature and climate change.
3. There is higher solar radiation.
4. The precipitation is also higher in tropical regions leading to the creation of more trophic levels and food chains.
5. Has more radiation.
6. An ecological balance is maintained with regard to a variety of pests and diseases.
7. Have ecological niches leading to higher biodiversity.
8. Larger outcrossing occurs leading to greater genetic diversity.
9. Higher amounts of soil organic matter which forms a strong base for maintaining healthy soil fertility.

The Effect of Altitudes on Biodiversity

It has been observed that when one goes up to an altitude of 1000 m, there is a decrease of 6.5 °C in ambient temperature, which will affect the biodiversity. Temperature in the mountain ranges goes on decreasing as one climbs up. There are several factors which contribute, like edaphic (soil related), high proportion of gravel, emergence of rocks etc. Mountain glaciers have lowest species biodiversity.

What Are the Values of Biodiversity?

Biological resources are constituted by genes, species etc. This is the foundation of life on planet Earth. The species are the foundations of Ecosystems. When one ponders the biological resources from a human perspective, one can elaborate on different values on a scale from ecocentric to anthropocentric. The starting point for an appropriate and sound attitude of humans and their institutions towards nature must be the foundation for ethical values of biodiversity.

The World Charter for Nature adopted in the General Assembly of the United Nations in 1982 recognizes that a part of Nature is humankind, and, that every form of life is unique and it warrants respect towards nature, regardless of it's utilitarian value to humankind. The manner in which, during the present day, man is instrumental in the destruction of this various forms of life, including soil resources, does not bode well for life and the future of life on planet Earth. McNeely et al. (1990) has proposed the following classification of biological resources:

A. Of Direct Value:

(1) Of Consumptive Use Value: Non market value of firewood, animal fodder, game and medicine
(2) Of Productive Use Value: Timber of commercial value, fish, crops

B. Of Indirect Value:

(1) Non Consumptive Use Value: Scientific research, recreation.
(2) Option Value (Value of maintaining options available for the future.
(3) Existence Value (Value of the ethical feeling of the existence of nature).

In the afore-said classification, one misses—as a direct or indirect value—the Life Support Value, which corresponds to the maintenance of the ecological sustenance process. The diversity of biological resources, be it genes, species, or ecosystems, provides the basis with which different human conditions can adapt to changes, and the basis for several social and cultural expressions as well (Shiva et al., 2005).

Which Are the Sources of Food?

The different biological resources form the foundation for all foods, be it for humankind or animalkind. A variety of foods, such as seeds, fruits, pods, buds, edible flowers, leaves, stems, roots (subterranean crops, Nair, 2023a, 2023b), mushrooms etc. are derived from plants. A variety of flowering plants produce honey, mediated by bees. Milk, also comes from a variety of fodder the cows consume. Fermented foods, such as curd, buttermilk etc., are prepared by the fermenting process in which microorganism are involved. Preparation of alcoholic beverages also involve microorganism for fermentation. An enterprising farmer in California makes beer out of turmeric. Cultivated plants are the source of nearly 85% of the foods we consume. As much as 75% of the total energy humankind spends is derived from just three carbohydrate-rich foods- grain crops—which belong to the grass family (Poaceae). These are wheat, rice and maize (corn). However, numerous varieties (genotypes) of these food grain crops are cultivated and consumed by humankind throughout the world. Among them are varieties of landraces and those developed using modern plant breeding techniques. The average life span of a crop variety is often short, 5–15 years, after which the variety begins to deteriorate. Hence, the regular development of newer varieties is an absolute necessity. Genes of the landraces or wild relatives of the food crops are vital to develop varieties laden with traits such as drought and pest tolerance/resistance. In this connection, the role of crop wild relatives for food security has very recently been highlighted (Nair, 2019b). Animals produce nearly 15% of human foods. New animal breeds developed by using superior genetic materials from native breeds help produce more milk, meat, eggs, wool etc. Not only the animal breeds serve as sources of food and fiber, but, also help carry out various agricultural operations like ploughing, carting etc.

Production of Drugs and Medicines

Various plants have been used to treat the different human ailments. China and India are the best examples. The traditional system of medicine, *Ayurveda*, in India, use innumerable plants and plant products to produce *Ayurvedic* medicines, which have extraordinary curative properties. It has been, more often than not, clinically proved that some of the *Ayurvedic* drugs are more efficient than allopathic or homeopathic ones. The most recent example is of using *Ashwagandha,* a medicinal plant, to treat COVID patients. *Ayurveda* is replete with the knowledge of about a variety of trees, shrubs, and herbs, of specific medicinal properties in treating specific human diseases. In the modern world, several pharmaceutical companies rely upon several plants with therapeutic values. The mushrooming of both nutraceuticals and pharmaceuticals is an ever expanding industry in countries, such as, India, China.

It is the diversity of plants and the derivatives (active principle/ingredient) from it that propel the pharmaceutical industry. Two notable examples emerge, first, quinine, and second, morphine. Quinine used in treating the global malaria disease is derived from *Chincona ledgeriana.* Morphine used as an analgesic drug is obtained from *Papaver somniferum.* The anti-cancer drug Taxol is derived from the bark of the yew tree, *Taxus baccata.* There are a large number of traditional drugs in the world derived from principles/ingredients in the plants.

Which Are the Cultural and Aesthetic Values?

India, the home of the Hindus, is the land where two important plants, namely, the Tulsi (*Ocimum sanctum*), and Peepal (*Ficus religiosa*), are extensively grown for use in religious ceremonies. Some wild, and some domesticated, animals, birds, and even snakes, are used in religious worship by the Hindus.

It is also important to note how biodiversity plays a key role in human existence. For instance, our human spirit enjoys nature's beauty. Eco-tourism is now a growing industry. A visit to wildlife sanctuaries, biosphere reserve, and places with aesthetic appeal allures people to see and enjoy nature's beauty, and, in reality nature's biodiversity.

What is the Existence Value?

All human and animal minds have feelings. The feelings of human beings could have many colors—positive, negative, vengeful, arrogant, despair etc. In fact, these feelings are a reaction of one's own thinking. However, the most important point to consider is that one should have only ethical feelings. Existence value means the value of ethical feelings about existence of nature. One often attaches value to many species even if one knows that one would never encounter them. We appreciate their existence and get fascinated by the very feeling that they exist in a specific environment or a habitat. It is important to remember that though many species and genetic resources may not be of direct benefit to human beings, they still are an integral part of nature, exist, as such. They also have a value for their existence on planet Earth. Their existence value often reflects in the sense of well-being. For instance, we know that biodiversity prospering around hydrothermal vents on ocean floor is of no direct benefit and there is hardly any desire to visit it. However, we derive benefit in terms of the knowledge of its existence and are fascinated by the very feeling of its existence. Until 1977, we knew that most marine life is dependent on photosynthesis. But, all that changed with the discovery of hydrothermal vents, especially of the giant tube worm. The chemosynthetic community of the hydrothermal vents, attributable to the existence value of this remarkable biodiversity, which contributes much to our understanding of biodiversity and is enrichment.

What is the Bequest Value?

One may dismiss a natural resource or a component of the same as inconsequential. Interestingly, the current generation might appreciate it to gift it to the future ones. This is "bequest value". This trait emanates from one's own mind that even if a current resource is of no value to the person concerned, presently/now, it might be beneficial to the future generation. Thus, non-use of a resource currently does not imply that it is of no use in the future. This shows that a perception which values biodiversity, based on current knowledge, is likely to attain a different perception in the future, when it is gifted or bequeathed to a future generation.

What is Option Value?

On planet Earth, there still are many species, in particular, of various plant species and microorganisms, which, as of today, are not ascribed any value. Often, one observes that we consider many plants as weeds and we attempt to eradicate them from the fields, using many means, mechanical and/or herbicidal. It is more likely than not, many of these plants and microorganisms, have some intrinsic value which we are unable to detect now, at the present. Hence, option value is the option of retaining this value which might be discovered at a later date in the future. It is quite likely, many of these plant species, which are now considered as weeds, might possess food value, which can be harnessed in the future. Also, many plants, which we consider as weeds now, might turn out to be the reservoirs of rare medicinal traits, of life-saving trait/value, which we might discover in the future. Some of these microorganisms which we consider harmful to day, might, in the future, turn out to be the sources for effective bioremediation of heavy metals, such as, Zinc, Cadmium, etc., and oil spills in the floor of the oceans and/or vital for the biodegradation of the so far non biodegradable substances.

What Are Ecosystem Services?

There are many functions that biodiversity performs such as the following:

1. Nutrient cycling
2. Sustaining food chains and food webs
3. Sustaining trophic levels
4. Sustaining energy flows

All of the above functions enable the self-sustaining nature of the ecosystem. Further, the formation, building up, protection and conservation of soil resources, conservation and purification of water resources, natural pest control, soil carbon sequestration, which is very vital to maintain soil fertility, especially in the face of high loss of soil carbon due to the chemically—centric extractive farming, euphemistically called the green revolution (Nair, 2023a, 2023b), maintenance of the gaseous composition of atmosphere optimally, climate regulation etc. Further, biodiversity infuses resilience and imparts a high degree of sustainability to an ecosystem, and, consequently to the entire biosphere.

Which are the functions biodiversity performs, which have an economic imperative? Singh (2009) catalogues the following functions:

A. Supply of goods, such as water, foods, fiber, fuel, and, medicines
B. The regulation of ecosystem processes, such as, quality of air, water and climate, soil erosion, manipulation of biological processes, risk reduction and cause of human and animal illnesses
C. Non-material benefits, which enhance enriching the quality of life such as cultural, religious and spiritual values, including eco-philosophy, knowledge inspiration, recreation and eco tourism etc.
D. Activities which are of back—nature/up support, such as, those required to produce other services, which include primary production, plant pollination, oxygen production, habitat creation, nutrient cycling, under the umbrella of soil resources

In sum, the term biodiversity implies, the huge variability that is present, among all living organisms, humankind, animal kind, and plants. These inhabit both the terrestrial and aquatic systems. Biodiversity prevails in three distinct hierarchical levels:

1. Genetic diversity
2. Species diversity
3. Community and ecosystem diversity

The chapter discusses, in detail, the myriad functions and inter-relationships among all the factors involved.

Variations in the natural biodiversity occur following the changes in latitude and altitude. Biodiversity gradually increases from high to low altitudes.

Some of the geographic locations of planet Earth are rich in biodiversity. For instance, India, is one of the countries in the world with megabiodiversity. In as much as India is concerned, there are ten biogeographical zones as follows:

1. Trans Himalayan Zone
2. Himalayan Zone
3. Desert Zone
4. Semi-Arid Zone
5. Western Ghats Zone
6. Deccan Peninsula Zone
7. The Gangetic Plains
8. North-East India
9. The Indian Islands
10. The Coastal Region

Biodiversity is of immense importance to mankind. More than 85% of all foods are produced from existing plant resources. In fact, biodiversity is the bedrock of sustenance for mankind and animal kind.

References

McNeely, J. A., Miller, K. R., Reid, W. V., Mittermeier, R. A., & Werner, T. B. (1990). *Conserving the world's biological diversity* (193 pp). IUCN, WRI, CI, WWW-US and the World Bank.

Nair, K. P. P. (2019a). *Intelligent soil management for sustainable agriculture—The nutrient buffer power concept*. Springer Nature.

Nair, K. P. P. (2019b). *Combating global warming—The role of crop wild relatives for food security*. Springer Nature.

Nair, K. P. P. (2023a). *Global commercial potential of subterranean crops—Agronomy and value addition*. Springer Nature.

Nair, K. P. P. (2023b). Extractive farming or bio farming? In *A better choice for 21st century*. Springer Nature.

Nautiyal, N., Singh V., Joshi, N. K., & Rastogi, A. (Eds.). (2012). *Climate change and biodiversity (228 pp)*. Biotech Books.

Shiva, V., Singh, V., Dankelman, I., Negi, B., & Singh, S. (2005). *Biodiversity, gender and technology in mountain agriculture. Glimpses of the Indian Central Himalayas*. New Delhi, Naddanya. 90pp.

Singh, V. (2009). Environmental services emanating from the Himalayan mountains: Valuation against the backdrop of eco-philosophy and chasing the goal of global happiness. In P. L. Gautam, V. Singh & U. Melkania (Eds.), *Ecosystem diversity and carbon sequestration: Climate change challenges and a way out for ushering in a sustainable future* (pp. 342–355). Daya Publishing House.

Soule, M. E., & Wilcox, B. A. (1980). *Conservation biology: An evolutionary-ecological perspective*. Sunderland, Massechutes: Sinauer Associates.

Wilson, E. O. (Ed.). (1988). Biodiversity. Washington, DC.: The National Academic Press. https://doi.org/10.17226/989.

Chapter 2
Agrobiodiversity

For the welfare of humankind, agro biodiversity is very vital. Agrobiodiversity meets all the food requirements of humankind. Not just this, even the fodder, for the animals, fuel and fiber for mankind, are requirements which are met by agro biodiversity. It is not just as a food requirement that human kind needs agro biodiversity. But, for the very meaning of life, where cultural, intellectual, ethical and aesthetic developments are met by agro biodiversity. It is not easy to comprehend this on the surface, but, a deep introspection reveals how, our very life depends on ago biodiversity. Fort a country' food security, agro biodiversity is critical. It also ensures the sovereignty of a nation. If one has to focus to ensure an adequate/sustainable future for a nation, it is inarguably important to protect, conserve and augment agro biodiversity. Table 2.1 details the classification of all living organisms according to their mode of nutrition.

It is through photoautotrophy that humankind feeds itself. This process is rooted in what we have come to call "agriculture". It is by managing specific biodiversity in the cultivated area that humankind manages its nourishment. The biodiversity sustained by photosynthesis is also the source for nourishment of animals. Agrobiodiversity is not only the exclusive source of nourishment, but, also, the basis for all kinds of livelihoods humankind design and develop. The affluence of agrobiodiversity during our present times is necessary for our sustenance and its continued preservation must be ensured in the interest of future generations.

What is the Difference Between Biodiversity and Agrobiodiversity?

While the term biodiversity encompasses agrobiodiversity, as well, agrobiodiversity, in itself, is an integral part of nature's biodiversity. All of agrobiodiversity is biodiversity, but, all of biodiversity is not agrobiodiversity, and, cannot be. Agrobiodiversity can be referred to as biodiversity, but, the entire biodiversity cannot be referred to as agrobiodiversity. In fact, the term agrobiodiversity is a, relatively recently, chosen

K. P. Nair, *Biodiversity in Agriculture*,
https://doi.org/10.1007/978-3-031-44252-0_2

Table 2.1 Classification of living kingdoms based on their mode of nutrition

Kingdom	Autotrophic (a plant making its own food)		Heterotrophic (a plant which cannot make it's own food/ parasitic)
	Photosynthetic	Chemosynthetic	
Prokaryotes (Bacteria and Archaea)	Yes	Yes	Yes
Protists	Yes	–	Yes
Plants	Yes	–	Yes
Fungi	–	–	Yes
Animals	–	–	Yes

one, or, a notion to describe selective biodiversity, managed and utilized by the vast humankind, to meet the various socioeconomic needs of the society, especially, the crucial requirements to meet the food, fiber, fuel, pharmaceutical, and, energy of the society, and numerous others of specific market value, through industrial processes, end products, what have come to be known as "value added products". For instance, fresh cocoa is an agrobiodiverse product, but, when someone drinks a cup of fresh chocolate made by Nestle, it becomes a value added product. Similarly, when corn is consumed as such, it is an agrobiodiverse product, but, when one buys a packet of Kellog's corn flakes, it becomes a value added product. Such examples can be vastly multiplied.

In fact, value addition to raw agrobiodiverse products, has opened up a huge international market, of mind boggling proportion/monetary value. The dark side of the entire marketing strategy is that a poor African farmer and his family from The Republic of Cameroon or Ghana, who cultivates, toils day and night to maintain a cocoa garden/plantation, gets only a pittance for his product, the company nestle makes colossal sums of money, running into billions of dollars through marketing its chocolate, the end product or value added product of cocoa. The sad part of this entire scenario is that almost all of these countries on the African and Asian continents, which produce many of these primary agrobiodiverse products, toil endlessly, and the European/American soil or climate cannot grow these crops, the entire pecuniary advantages go to the latter, at the expense of the former. This is part of the Liberalization, Privatization and Globalization (LPG) strategy invented by the Americans/ Europeans. And, the poor of the developing or never developing (not allowed to develop, as I perceive it) countries, stand to lose heavily.

Traditional Agriculture

The traditional agriculture had a larger cultivated area, and had strong linkages with uncultivated (forest/grassland/rangeland) areas. Since these poor farmers who tilled, included a large number of cultivable annual plants, largely food grain crops, such as, wheat, rice and millets etc., they depended on a very broad genetic base of many foodgrain crops. Genetic diversity in traditional farming was a unique example of the farmers' art of cultivating agrobiodiversity. Many species of livestock were also reared. Draught animal power and human muscle power were only the sources of energy (animate/muscle energy) to operate field ploughing, carting agricultural produce etc. For making the bulk of human foods, one requires fuel for food processing, and, wood was the only source of fuel for this. Hence, traditional agriculture inflicted huge pressure on forest land. In rural areas, the bartering system was also present. Since there was broad-based agrobiodiversity, traditional agriculture was also sustainable, in its own abilities and potential.

The "Green Revolution" Agriculture—What Are Its Implications?

It was Dr. William S. Goud, Director of the United States Agency For International Development (USAID) who first coined the term on March 8, 1968. Since then, it has stuck, in global dialogue on the subject of farming. Before one goes further on the discussion, one must understand some basic aspects of this type of agriculture. The following are some key points:

1. First and foremost, the very term "green revolution" is an euphemism. It was never "green" and never it will be.
2. This type of agriculture is characterized by the following key points:
   A. Sowing the dwarf high yielding wheat and rice varieties, which were then popularly known as the "miracle" and "High Yielding Varieties" (HYV) or "High Responsive Varieties" (HRV)—highly responsive to chemical fertilizer inputs (though this author found that it was only for a short period, before these HRVs broke down in their capacity to yield more). It is important to know that all of these carried the alien blood
   B. These HYVs or HRVs required high amounts of chemical inputs, primarily synthetic fertilizers, herbicides (to control the weed population) and pesticides (to control the disease and insect pests)
   C. High amount of irrigation water was required.
3. In other words this type of agriculture was a highly soil-extractive agriculture based on heavy inputs (Nair, 2023b).

What Were the Consequences?

The green revolution held its sway for about a decade, at the most, between mid sixties and mid seventies, primarily in South Asia, in particular, India, by which time, the following consequences were noticed.

1. Agrobiodiversity had practically vanished due to continuous monoculture of wheat and rice, though, it must be clearly admitted that a mountain of rice and wheat grains was produced during the period, when the green revolution held it's sway. This was true of the State of Punjab, and Western Uttar Pradesh in India, in particular, and, Punjab, was called the "cradle" of Indian green revolution.
2. Soils, especially in Punjab State and Western Uttar Pradesh State, where green revolution held it's sway, were completely degraded. The carbon levels in soil sunk steeply, ground water was polluted with chemical residues of synthetic fertilizers and pesticides, and the water was no more potable. Of India's total 328.73 million hectares of geographical area, as much as 120.40 million hectares became degraded as a consequence of green revolution.
3. Districts like Gurdaspur in Punjab State, came to be called as "cancer capital" of India, due to unbridled use of synthetic insecticides and fungicides, residues of which found their way into the human body causing cancer.
4. There was huge spread of plant diseases like wheat rust and the rice crop in Asia, practically, being decimated due to the attack of Brown Plant Hopper (*Nilaparvata lugens* Stal.) insect. The heavy dosage of insecticides decimated the natural predators/enemies of BPH and BPH population increased uncontrollably, totally decimating the rice crop in Asia.
5. Global warming increased by about 30–35% due to the emission of nitrous oxide from unbridled use of urea (Nair, 2019a, 2019b).

The bottom line of green revolution:

Though green revolution produced a huge amount of food grains, especially of rice and wheat, it left a trail of huge environmental disasters. These aspects are thoroughly discussed by Nair (2019a, 2019b) and an alternative and preferable way of farming is proposed by hm in a forthcoming book (Nair, 2023b). Green revolution was simply market-oriented.

What is the LPG Agriculture?

The term LPG stands for "Liberalization, "Privatization" and "Globalization", respectively. This type of agriculture was ushered in towards the end of the 20th century. And, this has transferred the power to private industry. The LPG syndrome is now omnipresent. Global agriculture is now in a transition phase, from the highly input—centered extractive farming, euphemistically called the green revolution, to the market controlled agriculture of the LPG. This agriculture is biotechnology driven, controlled by the global market forces. This agriculture is largely meant for the profits of the corporate sector, mainly those in seed business, like Monsanto and Bayer. New crop cultivars, that nature's evolution "failed" to add to nature, have been created by genetic engineering. Genetically modified organisms (GMOs) belonging to crop cultivars and animals are emerging in laboratories, financially controlled by multi national companies (MNCs). The fate of the Bt crops, of which the Monsanto's Bollgard cotton in India is one eye opening example. Touted as a solution to the dreaded pink bollworm attack of cotton, this Bt cotton was sold to Indian farmers at a huge price. This author had predicted its failure, in course of time, warning that the pest would mutate and new infestation would emerge. This happened within three years of extensive cultivation of Bollgard cotton in northern India. Thousands of cotton farmers committed suicide when the Bt cotton crop failed miserably due to new pest attack, in the cotton belt of Maharashtra State, in the Vidharbha region. This mass suicide, due to indebtedness (the farmers faced very huge financial losses), a blot on Indian agriculture. Never before in the history of agriculture in the world, did occur such mass farmers' suicide, as in India. Finally, Monsanto had to pack up and leave India, for ever. Some vested interests refer to this as the "Gene Revolution"—an extension of "Green Revolution".

A similar thing like the foregoing example is unfolding in India, currently, with the genetically modified mustard crop, called the "Dhara Mustard Hybrid" (DMH-11). Mustard crop cultivation is primarily confined to the North Indian belt and (DMH-11) is developed by a Delhi-based scientist. This author finds the recommendation of DMH-11 for India wrong on two counts: Firstly, the mustard crop employs a very huge women labor force for weed eradication in mustard crop. Secondly, the recommendation is to use Glufosinate Ammonium (GA) for weed control. GA is suspected to be carcinogenic. It's use will displace millions of women laborers job, rendering them jobless, leading to pecuniary problems. On both counts, this author finds the recommendation to use DMH-11 neither scientific nor economically beneficial for India. The private company, involved in the in collaboration, is pushing for DMH-11 adoption in India, just like that in the case of Monsanto's Bollgard cotton.

Both agrobiodiversity and the environment are completely ignored in LPG agriculture. This development will have far reaching implications for the environment, ecosystems, and public health. Wen occupying a significant proportion of cultivated land, the Bt cops would completely wipe out biodiversity, of insects, including those playing the role of pollinators. This would ultimately culminate in the breaking down of ecological integrity and extinction of several species of pollinators and other insects, many of the other crop cultivars, and, livestock breeds. Agriculture is intelligent anthropocentric management. Nature's biodiversity, squeezed out as agrobiodiversity, by being brought into cultivation practices, began with the beginning of land tillage in the primitive stage of agriculture. In traditional agriculture, people were the masters of agrobiodiversity, which in fact, is a "public property". In the green revolution phase, agrobiodiversity slipped into the hands of private market, especially the seed manufacturers, who are profit-oriented, resulting in the establishment of several seed and fertilizer manufacturing companies, of which, Monsanto is the leading one (at the time of writing this book, Monsanto has been bought over by

Bayer). In the modern LPG era, agrobiodiversity is controlled by vested interests, which has the capital, and availability of human food is controlled by the private market system.

The global natural resources have been severely affected by the indiscriminate and improper management of the anthropocentric biodiversity/agrobiodiversity. While primitive agriculture had its ecological roots and was represented as the most sustainable system of food production, in the entire history of agriculture, the 20th century highly chemical-centric extractive farming, euphemistically called the "green revolution", and, it's successor, the "gene revolution", which were at the center of human greed, has completely upset the foundation of sustainable agriculture, in fact, sustainable human and animal life. The unsustainability of modern agriculture arises out of the total destruction biodiversity, in particular, agrobiodiversity. The Industrial Age, that began in the 1750s, with the invention of the steam engine for locomotion, had many things and processes in its "Pandora's box". One of them is the gene-led industrial agriculture, within the global umbrella of the LPG agriculture, that came in about 250 years after the beginning of the Industrial Age, and, the Industrial Age is all set to assume its climax. The invention of the Internet, and its extension, the Artificial Intelligence, does not bode well for humanity, as pointed by Nair (2023a) in his most recent book "The Living Soil". Industrial agriculture, that is, the green revolution and the gene revolution, are slowly, but steadily, engulfing the whole system on which agriculture is based, the sustenance of humanity, and, on which all hopes of life survive (Singh et al., 2014).

The Biodiversity—Food Production Link

The living organisms, be of plant origin or animal origin, provide all foods. One species provides food for another. The greater the degree of biodiversity, available in the case of a specific species, the higher the amount of food it can provide in nature, for other species. Natural evolution determines the choice of a living organism for its food needs. Inasmuch as human beings are concerned, they choose from a great variety of plants, such as herbs, shrubs and trees, animals (aquatic and terrestrial, vertebrates ad invertebrates, insects and reptiles, avian species and mammals), and, even microorganism (bacteria and fungi), and, humans, relish a great variety of foods. Inasmuch as humans are concerned, there is no kingdom, no family, no phylum, and no species from which humans do not choose their foods.

Agrobiodiversity, depends on the following factors:

1. Geographical or agro climatic zones
2. Religion, culture and traditions
3. Market value and demands

The most important factor, geography, is linked to agroclimatic and agroecological specifications. And, this determines the nature of plants which produce foods, which thrive and can easily be cultivated in a specific geographic region. However, there are quite a number of plants, which are not among the native ones, which are being cultivated throughout the world. One can find such examples from many regions of the world. For instance, corn or maize, was introduced into India by the Portuguese in the 16th century from the Mexican region of the Andeans. It originated from Central America and Mexico. It now thrives very well in the country (India). The popular Dragon Fruit (*Rambuttan*) introduced from China, is now, very popular in India, especially, in States like Kerala. Examples can be multiplied. Other major factors which play a crucial role are:

1. People's religion
2. Culture
3. Traditions

To elaborate, mostly, the Hindus of India are vegetarians. Hence, they cultivate those plants which produce foods and fulfil their nutritional requirements. Many other tribes in the world depend on forests for their food. Many religious societies choose some specific plants for conservation as per their cherished values. Some cultural groups are found to grow specific native plants which provide foods, which are specifically used on special occasions, festivals, rituals etc. The Hindus, in India, celebrate some sacred notions in their culture, such as, sacred forests, groves, sacred trees, sacred seeds, sacred animals etc. These are pivotal in biodiversity and agro-biodiversity conservation. It is also important to note that market considerations also propelled people to cultivate crops, enabling agrobiodiversity to thrive. This helps both native and introduced crops to be widely cultivated in different regions. In the market driven economy, farmers prefer to cultivate the improved crop cultivars rather than native ones at the expense of agrobiodiversity.

For food security, and food sovereignty, biodiversity is the foundation. The broader the base of biodiversity, the higher the level of food security and food sovereignty. A large proportion of biodiversity-rich habitats has been reduced to considerably small habitats, which are attributable to indiscriminate and callous deforestation, for converting biodiversity-rich lands into cultivated lands. The plunder of the Amazon forest is a classical example. The cultivated lands support only what we term as agrobiodiversity—that is cultivating an array of food, fodder or fiber crops, and, not Nature's evolution-led biodiversity. Being reduced to just a few food yielding and commercial species, with extremely narrow genetic base of each species, the cultivated lands have pushed Nature's biodiversity to almost extinction. A new Report by the World Wildlife Fund has revealed that consumptive agriculture-dominated wheat, rice and maize, is wrecking havoc on wildlife conservation. Not only wildlife conservation, but, these three crops are also threatening the entire food security web of the entire world. Most of the destruction of biodiversity in agriculture has also happened because of these three crops, which are like death knells for Nature and food security. As per the projections of David Edwards of the World Wildlife Fund, wildlife populations have declined by 60% since 1970 due to global agriculture (Singh, 2022).

More than 75% of the food human populations consume, is derived from twelve plants and five animal sources, with about 60% of the calorific entitlement in the entire human diet originating from these three crops—wheat, rice and corn. In fact, it is akin to a stranglehold. Strangely, palate decides the consumptive pattern. For instance, if one takes into consideration the Asian continent, it is rice. Late Richard Nion, President of the US, once famously remarked that those who control rice production, will control the destiny of the Asian population. This was a precursor to the starting of the International Rice Research Institute in Los Banos, Philippines, ably financially supported by the American Ford and Rockefeller Foundations. Their stranglehold on Asian agriculture emanated from this thought. As per the estimate of the FAO (Food and Agriculture Organization of the United Nations), 51% of the calories consumed by the global population originated from wheat, rice and corn. Billions of people around the globe rely entirely on these three crops solely, for survival. A disruption in their production will have tragic consequences for human civilization. And, interestingly, it is the Chicago market that controls wheat price, globally, not Indian. One can visualize how the LPG syndrome holds the world to ransom.

The World Wild Life Fund in its report titled "Future 50 Foods for healthy People and a Healthy Planet" (Knorr and WWF, 2019), has concluded that the conservation of wildlife, and, even human survival, is dependent on the way humans eat the food. When the same crop is planted in the same area, year after year (the monoculture practice, for example rice–wheat rotation, of the extractive farming called euphemistically the green revolution), it leads to the steep carbon depletion, resulting in the loss of inherent soil fertility, and other nutrients, as well. Such crops demand excessive use of synthetic fertilizers, pesticides and herbicides, which are deleterious to the inherent soil fertility. These aspects are thoroughly discussed in the book "Intelligent Soil Management For Sustainable Agriculture—The Nutrient Buffer Power Concept" (Nair, 2019a).

Consequent to modern agriculture, there has been a staggering decline in wildlife species during the last half a century. Insect species have been decimated in many places, which has very adversely affected pollination. Millions of plants will be extinct during the coming decades and food security will be seriously threatened. In traditional farming, there used to be thousands of varieties in the three crops, rice, wheat and corn, on which green revolution focused, but, they all have disappeared consequent to the green revolution. For example, in rice alone, there were about 60,000 varieties in India, but, consequent to the green revolution, only a handful are now being cultivated.

In the world, while the European Union and China produce about 20 and 18% of the total wheat produced in the world, India's contribution is about 13%. India, China, Bangladesh, Indonesia and Vietnam top the global rice production, of which China tops at 24%, while India is at 19%. The US, China and Brazil account for 32%, 22% and 10% of the global maize production. The US is the largest exporter of maize in the world. It is very important to note that about 75% of biodiversity has been lost in the agricultural sector according to the FAO. In Britain, there are about 24,000 varieties of apple, but, sadly, only about 20 varieties only can be now found in the supermarkets.

In our present day world, agriculture is no more aimed at ensuring food security, directly, and, it is controlled by the demands of the market. Sadly, what the market wants will be grown in the fields, not what people want! The key to food security is also in the hands of market. Destroying the vast diversity of the food providing plants, and giving supremacy to only just three crops, namely, wheat, rice and maize, is the most bizzare contribution of the green revolution and LPG agriculture.

The most critical need of the preset moment in the history of mankind, is to adapt our agriculture according to the principles of environmental integrity and transform food production according to the socio-cultural needs of humanity, which encompasses, physical, intellectual, moral, ethical, emotional and psychological needs. It is the artificially imposed uniformity, to tailor to the market demands, that has led to global food consumption restricted to just three of the Poaceae grass family, namely, wheat, rice and maize. Unless this stranglehold is broken humanity will witness the complete disappearance of agrobiodiversity.

It is important to note that while the chemically—centric highly soil extractive farming, euphemistically known as the green revolution, set in motion the destruction of agrobiodiversity, centering simply on monoculture, around three crops, namely, wheat, rice and maize, the LPG agriculture will spell ecological disintegration, leading to complete wipeout of genetic diversity, with the genetically modified crops, such as Bt crops (example, Monsanto's Bollgard cotton). Transgenic crops will lead to the complete destruction of a variety of insects, including, the most beneficial pollinators.

What is the Role of Agriculture in a Sustainable Future?

It is important to remember that agriculture did not evolve suddenly. It has taken millennia to evolve. Different systems of agriculture evolved in different geographical regions of the world, under different cultural settings. Primitive agriculture and traditional agriculture encompassed an enormous amount of biodiversity of farming systems. Hence, agriculture under these two systems is extremely resilient and sustainable, which ensures a sustainable future.

It is important to realize that a sustainable future must ensure two very important conditions as follows:

1. A happy humanity
2. A blossoming humanity

The above two criteria must ensure the following:

1. Availability of plentiful foods for humanity
2. Ecological affluence

Availability of plentiful foods implies a healthy, vibrant, environmentally safe, ecologically ameliorating, resilient, and sustainable agriculture. Ecological affluence implies natural resources in a healthy and regenerative state, biodiversity at its climax, and human beings in a healthy relationship with all life—plants and animals. It is important to remember that availability of plentiful foods and ecological affluence are bound to each other, and are, inseparable. Ecological affluence ensures plentiful foods obtainable from innumerable plants, both cultivated and uncultivated. It is ensured by pursuing, following and sustaining the most appropriate food production systems. And, a sustainable future is ensured by a healthy, and, ameliorating relationship between the two.

The experience of the recent past, especially after the global implementation of the green revolution, clearly establishes the fact that green revolution and LPG agriculture cannot and will not take humanity to a sustainable state. This is because, both are based on a severely shrunk agrobiodiversity, and they are severely polluting, to both the soil resources and the environment (Nair, 2019a, 2019b), degenerative and disastrous (Nair, 2023a, 2023b). On the other hand, the other two categories of agriculture, primitive agriculture and traditional agriculture, are life-enhancing, and, therefore, ensures that humanity ushers in a happy and sustainable future.

A Pyramidal Concept of Agriculture *Vis-á-Vis* the Future of Humanity

It is important to know and recognize that there still are some geographical regions, in the world, and specific cultures, which still operate primitive and traditional agriculture, notwithstanding the onslaught of the chemically-centric green revolution and LPG agriculture, elsewhere. Mountain ranges of the Himalayas, tribals, in almost all Indian states, and many other regions, and tribes in African continent and Asian continent still maintain their identity with primitive and traditional agricultural systems. There still are numerous tribes in the world, who do not plow their fields, for growing crops, but, pick up foods from edible mushrooms, roots, shoots, leaves, buds, flowers, pods, fruits, honey etc. These are found aplenty in natural forests which they inhabit. There still are areas where farmers plow the land using draught animal power. They continuously enrich their soils with organic fertilizers, like farm yard manure (FYM), green leaves (from plants, such as Sunhemp, *Crotalaria juncea*, or *Glyricidia maculata*), compost, etc., and, cultivate a large number of landraces of all food crops, such as, millets, cereals, pulses, oilseeds, fruits, vegetables, condiments, and, spices. These farmers only use traditional methods of plant protection, and, avoid all sorts of chemicals, both to control weed and insects and fungal and bacterial pests. Their approach to food production is farming system—based and they maintain an appropriate ratio between uncultivated (forests/rangelands/grazing lands) and cultivated lands (rear many livestock, such as cattle, buffalos, sheep, goats, chicken etc.) as part of their farming systems and farming cultures.

The above described farming systems are now giving way to the imposition of both the green revolution method, where the emphasis is on high yielding varieties

(HYVs) and high responsive varieties (HRVs), both being market-driven. Additionally, comes the pressure of the LPG agriculture, where the emphasis is on transgenic varieties. There is huge institutional and/or governmental pressure on these farmers to switch from the traditional agriculture, which they have been practicing since generations, to these new ideas, which, in the short run, will be hugely economically profitable, but, in the long, ecologically disastrous. As we have already seen in the previous discussion, since most of the photosynthetic energy, converted to nutrients, which nourishes humanity, is at the base of the ecological pyramid structure, the primitive or traditional agriculture would support the largest human population. However, this balance is now being seriously threatened by the new market initiatives of the green revolution, which one might even call "greed revolution" and the gene revolution. In an ecological pyramidal structure, the primitive/traditional agriculture would support the largest human population. Hence, the future of humanity is safest based on this agriculture. However, the market-driven economy, the bedrock of the LPG agriculture—Liberalization, Privatization and Globalization concept—is planned to demolish the earlier structures. In fact, the chief aim of the market economy is to concentrate wealth in the hands of the chosen few. To illustrate, for example, seed production (both hybrid and transgenic) would be controlled by multinationals like Monsanto/Bayer. Fertilizer production would be controlled by other multinationals (MNCs). The seed-fertilizer combine will see to it that the entire base of traditional agriculture is uprooted. In fact, agriculture will be completely controlled by the market.

An analogous discussion would be in order in this context. Thus, the green revolution and LPG agriculture behave like carnivores, in the ecological pyramid, which will consume much of agrobiodiversity of traditional agriculture, thus limiting the potential of photosynthesis, the foods—producing plants perform. Hence, energy and nutrient levels are considerably reduced at the green revolution level, and, it will support only limited population, or large population for a limited period only, as has happened in India, consequent to the green revolution, which produced a lot of food in the short run, but, both agrobiodiversity and soil biodiversity were seriously affected will totally vanish. LPG agriculture behaves like top carnivores, which has devoured (or will eventually do so), most of agrobiodiversity, replacing it with fewer transgenic crops. This trophic level lying at the tip of the ecological pyramid can be equated with top carnivores, like a lion, tiger, cheetah, vulture etc., who are not left with the choice of varieties of plentiful foods to consume. Hence, many top carnivores are at the verge of extinction. One can visualize the fate of humanity if it has to depend on LPG agriculture—it will be total doomsday!

Mountain Agroecosystem—What Are Its Characteristics/ Specificities?

The mountain regions have distinct specificities, as compared to the plains. Jodha (1990) has dealt with it. Mountain regions have the following six specific characteristics:

1. Limited accessibility
2. Marginality
3. Fragility
4. Diversity or heterogeneity
5. Ecological niches or comparative advantage
6. Human adaptation mechanisms

Of the above, the first three characteristics are the negative attributes of the mountains. The last three are the positive attributes. The above-mentioned mountain ecosystems specificities determine the level of biodiversity in the region. They are as follows (Jodha, 1990; Jodha et al., 1992).

1. Limited Accessibility: The mountain slope, altitude, conditions of the terrain overall, also periodical seasonal hazards, such as, landslides, snowfall, avalanches etc.).
2. Marginality: This means the characteristic of the mountains that counts the least concerning "mainstream" situation, and, one of the most commonly known mountain specificity. The basic factors which contribute to the marginality of the mountains are remoteness/physical isolation, fragile or low-productivity resources, and, several man-made handicaps.
3. Fragility: It is attributable to altitude and steep slopes, in association with geologic, edaphic, and, biotic factors, which limit the capacity of the mountains to withstand even a small degree of disturbance. When the general environment and the resources of the mountains begin to deteriorate or vanish, due to any of the disturbance, the effects of fragility occur at a much faster rate.
4. Heterogeneity or Diversity: The extreme degree of this attribute is a function of the interactions of the various factors, such as, elevation, altitude, geologic and edaphic conditions, steepness and orientation of slopes, wind and precipitation, mountain mass, and the relief of the terrain.
5. The Ecological Niches: The comparative advantage of the mountains over plains is revealed by these factors, which are specific to their environmental and resource-related features, due to which mountains provide a "niche" for specific activities or products unique to mountain ecosystems.
6. The mechanisms of adaptation: The mountain communities have successfully evolved these mechanisms, through trial and error, over a generational space. The formal and informal arrangements to manage mountain resources and diversified interlinked activities clearly reflect the adaptational mechanisms or experiences.

Comparison of Mountain Agriculture and Mainstream Agriculture

Among the global mountains, the Himalayas represent the tallest, youngest and most fragile mountain, on the planet Earth. It is a unique geological formation. There are ever so many religious legends (especially Hindu) surrounding it. Himalayas is the abode of the Hindu Deities, especially, Shiva and His consort Parvati. Fragile mountain ecosystems are found to possess a high intensity of biodiversity, namely, alpha, beta, gamma. The Himalayan mountain ranges play a very interesting role unmatched with what one encounters in the plains. While global agriculture has gone through massive changes, which includes those in the agrobiodiversity patterns, it is educative to note that farmers in the Himalayan mountain ranges have not drastically changed their farming systems, as compared to those in the plains. When one critically assesses the high degree of inaccessibility, marginality and fragility of the ecosystem, there is, but, little scope to hope significant changes in farming landscape in the Himalayas. To a great extent, mountain agriculture still retains its traditional features. Cropping systems tend to ensure agrobiodiversity following farming systems, as for example, altitude, slope, irrigation facilities etc. Table 2.2 catalogues some of the important/ contrasting features of mainstream agriculture versus mountain agriculture.

Diversity of Agroecosystem

The field-related agriculture is often cut off from the Nature and its components. Agriculture, in our general understanding, concerns all land-related activities, including cropping, forestry (more related to the later development of agroforestry, rather than general forestry), horticulture, olericulture, seed production, animal husbandry etc. An agroecosystem includes croplands, forests, grass lands, rangelands, livestock, water resources etc. The components of Nature which do not directly play a part in crop production, such as forests, grassland, rangeland or a water resource, that are not used for farming culture directly or indirectly, are not a part of an agroecosystem. In fact, an agroecosystem is an arbitrarily defined unit of Nature, with distinct energy flows and chemical cycling, involving uncultivated land areas, such as forests, grasslands, grazing lands, rangelands, cultivated land, and, livestock linkages with each other, woven finely into a complex unitary whole, functionally oriented to produce foods and other life-supporting products, such as feed, fodder, fiber, fuel, fertilizers etc., while maintaining the ecological equilibrium. An agroecosystem overlaps diverse ecosystems designed and managed by a farming community for the production of fundamental life-sustaining products. A core component of the whole biodiversity is the uncultivated lands—forests, grasslands and rangelands—which nourishes the cultivated lands, through constant nutrient cycling. Cultivated land is managed through ethno-engineering, often by clearing original forests. Livestock are the dynamic component of an agroecosystem: nutrient flows from the uncultivated

Table 2.2 Distinguishing features of mainstream agriculture and mountain agriculture

Distinguishing feature	Mountain agriculture	Mainstream agriculture
Fragility	Can be widely observed	Moderate
Marginality	Very high	Low
Inaccessibility	Very high	Totally absent
Diversity/heterogenity	Very high	Minimal
Niche/comparative advantage	Very high	Very poor
Adaptation capability	Very high	Very poor
Complexity	Very high	Minimal
Vulnerability	High	Moderate to high
Resilience	Very high	Very low
Water use efficiency of crops	High	Very poor
External inputs	Internal	External
Productivity	Moderate	Moderate to high
Farming system	Present	Haphazard*
Sustainability	Very high	Very poor

Note *The farming system is market controlled, note the monoculture system, (continuous rice–wheat rotation) in particular, of Punjab State, India, the "cradle" of Indian green revolution, which has been market controlled (the incentive of high price or what is known as the "Minimum Support Price" offered by the Government of India), which has ruined soil productivity of the State in course of time. The monoculture is no more sustainable because of huge soil degradation, pollution of ground water and steep depletion of the same, contamination with fertilizer and pesticide residues, leading to making the water no more potable
Source Jodha (1990), Jodha et al. (1992), Singh (2018)

land are mediated by the livestock. A community of farmers living in a cluster of houses, called a "village", manages an agroecosystem.

Ecological stability is ensured by uncultivated land where there is a flourishing biodiversity. A forest also enhances resilience, complexity, progressive succession, ecosystem efficiency, productivity and, eventually, sustainability of the farming system.

The agroecosystem should be regarded as a foundation for conceptual and practical approaches to the sustainable development of agriculture. The uncultivated land—cropland ratio is so very crucial for ecological sustainability, of the agroecosystem. It should be decided as per the agroecological zone in which an agroecosystem is located. Not more than two-thirds of the cultivated land in the plains and one-third in the mountain range should constitute the agroecosystem. Such a unit will be ecologically sound in vulnerable to biotic and abiotic stresses. Thus, biodiversity in the uncultivated and cultivated components would thus be saved, conserved and enhanced. Plentiful nutrient supply, from the uncultivated land, to the cropland would ensure the self-containment features of the farming system. Also, atmospheric moisture circulation and water conservation within the soils of an agroecosystem, would, thus be ensured.

The cropland soils is constantly enriched by a biodiversity-rich mountain agroecosystem, through continuous internal nutrient cycling. This nutrient cycling supports a healthy balance of soil organisms, including, bacteria, fungi, actinomycetes, protozoa, millipedes, insects, and, a host of other soil-bound creatures, which are essential to sustaining soil health (Nair, 2023a). This has been clearly explained in the book The Living Soil (Nair, 2023a). The phenomenon enhances high soil productivity. In an agroecosystem approach to food production, the soil is not regarded as an inanimate substratum, but, a living and thriving system (Nair, 2023a). The productivity of a system is a function of soil ecosystem. A soil ecosystem reverberating with a bounty of life forms, such as, microflora, microfauna, and macrofauna, will have a great bearing on the health of life forms in the troposphere, including that of plants, humankind and other animals, thriving on this substratum. A lot more food would be produced from such a system, than would have been produced from merely the cropland, for a variety of foods would be available from the uncultivated areas also. The system would be highly profitable for the enterprising farmer, because, he/she would earn from the produce of not merely the crop land, but also, from the uncultivated area. Demand for external inputs from the market would be reduced, and, in the long term, totally eliminated. Here, of course, would be no agriculture-borne pollution of the environment. Sound health of the system, livestock and the inhabiting people would be ensured. The system would also promote socio-cultural cohesion, amidst the community, because, the uncultivated area, if needed, could be used as common property resources (CPR) of the inhabitants. The ecological integrity of the farming system would be maintained and the sustainability of agriculture upheld.

The Uncultivated Land

In the mountain range, most of the land is uncultivated. These cover forests, rangelands, gazing land, grassland, marginal land, abandoned cultivated land, open spaces, homesteads etc. In the case of fragile mountains, the ecological stability of the system, should be the most important criterion for agricultural sustainability (Singh, 2005). The natural forests, harboring a great deal of biodiversity make an agroecosystem more vibrant functionally, thus, more efficient. Forests also act as the foundation of sustainability of the whole agroecosystem.

With regard to an agroecosystem, the contribution from forests could be measured quantitatively, while it may also be invisible. In as much as the former is concerned, wood, timber, fiber, fruits etc., are the direct quantitative benefits/products. Further, they also conserve soil moisture, and, add to nutrient cycling, through leaf fall. The carbon balance is very positively affected, which is a vital criterion n maintaining soil fertility.

For rural populations, fuel wood is the most important source of energy. Many activities, primarily cooking, is done by using firewood. This is typical of India and Africa. Alternative energy sources, as for instance, cooking gas, are unavailable in rural areas. Additionally, villagers in India also dry cow dung and use it as fuel.

Conventional agriculture in the plains demands constant sources of energy. Fossil fuel is the most important source of energy in this area. In the mountain agroecosystem, mechanization is totally absent, and fossil fuel is hardly in demand. The need for soil nutrients is met by using farm yard manure (FYM) and the nutrient recycling management, mediated by livestock.

Ecological farming in the fragile mountains should be the foremost criterion for maintaining sustainability. The crop and livestock are supplied with massive amounts of nutrients in a forest. Also, a forest plays a key role in the hydrological aspects. Thus, a forest continues to be a key feature of mountain agriculture. Deep rooted woody perennials contribute much to the organic pool of the soil through leaf shedding. The organic matter thus accrued in the soil enhances the soil nutrient status, water retention, and, also improves soil structure. The deep rooted trees also act as wind breakers, also cutting down evapotranspiration. Leguminous perennials will add nitrogen to the soil. Trees play a crucial role in resource intensification centered sustainable land use (Jodha et al., 1992).

Trees add much vitality to the entire agroecosystem. Since renewable plant diversity and recycled nutrients in the self-contained mountain farming systems are the basis of soil fertility management, no pollution in the system is caused because eco-malignant pesticides are totally avoided. Hence, in sum, a forest, or an assemblage of a community of trees, provides a healthy base for ecologically sound and environmentally safe agriculture system. Forest management practices are those which involve the use of the produce of the ecosystem, such as forest biomass, which can be used as feed for the livestock and as manure for the cropland. Hence, the forest acts as a linkage amongst the various components of the farming system. This contribution is vital to maintain the ecological integrity of the whole system. Forests affect so many of the important functions, such as:

1. Positive role in biodiversity maintenance
2. Helps natural regeneration of vegetation
3. Nutrient flow and recycling
4. Moisture circulation
5. Building and binding Soils
6. Vital for infusing sustainability into the whole system

The Cultivated Land

It is very important to note that the ratio between uncultivated and cultivated land is too wide to be ignored. If one takes the example of the Uttarakhand State, in India, (north India), only about 10% of the land is cultivated. The crops are cultivated on terraces, as it is very difficult to cultivate on slopes, some of which can be very steep. To raise food crops, terracing has to be done. Minimum terracing is an absolute must to raise food crops in the mountain range. This is because the mountain ecosystem is very fragile. However, in some mountain ranges, intense terracing is done, and, this is not advisable since the mountain ecosystem is very fragile. Terraces on gentle slopes, by traditional farmers of the mountain range, is a brilliant form of ethno-engineering. Almost all terracing is done by manual labor and/or draft animals used for carrying loads of goods across mountains, which mostly are mules/donkeys. The field operations are ploughing, puddling, interculture, threshing of produce etc. Women labor is also used, not men labor.

The cultivated land can be classified into two categories:

1. Irrigated
2. Rainfed

In the valleys, the cultivated land is almost plain. In the uplands, mid or lesser Himalayas, high altitudes and greater Himalayas, the cultivated and is largely rainfed. The rainfed terraced fields are used for raising crops without irrigation, and the cultivation practices in these area, invariably involve agroforestry. Cultivated land in different microclimatic zones raises different crops of different varieties or landraces, employ different cropping patterns. The cultivated—uncultivated land linkages are maintained naturally, as for example, nutrients from forests to cultivated land flow through the medium of rain, wind, or by human and/or livestock intervention.

The Livestock Scenario

It is very important to recognize the important fact that, in the mountain range, livestock constitute an important component of life. This is because, animals, like bullocks, mules, can perform tasks, which human labor cannot. The livestock form an integral part of the mountain farming system and a "lively bridge" connecting the two types of land, namely, the uncultivated forest land and cultivated land in the plains. It is very important to recognize the basic and crucial fact that this "lively bridge/linkage" maintains the healthy interconnectedness between ecological and socio-economic/cultural sustainability of the entire mountain ecosystem. Forests, especially the natural ones, are a very rich repository of many important plant nutrients, especially, they are "rich carbon banks", which hugely complements the cultivated lands in enriching their fertility. These nutrients are transferred to the cultivated land via the livestock. The nutrient transfer takes place in the following manner:

1. Forest biomass—tree leaves and ground flora—are fed to livestock.
2. Biomass is also used as a bedding material in livestock sheds. Both the dung and bedding material are converted into manure through composting which is then transferred to the cultivated field. This manure, is highly soil fertility enriching. Livestock also recycle the nutrients in the cultivated land. Crop residues are fed to the animals and thus, the nutrients in them are recycled into the cropland through their dung.

It is very important to note that the biomass transfer and cyclic flow pattern of nutrients mediated through livestock infuse vitality in the production system and also livestock which themselves fulfill their requirement to produce milk and provide muscle power for draught. This interlinkage between the forest ecosystem, livestock, and cultivated land, as already mentioned, is vital for the sustainability of entire mountain agriculture. The livestock are involved in the flow path for energy capture for both milk production and provision of draught power. Animals might not appear of direct use to the system, inasmuch as nutrient capture is concerned through energy mobilization, because, plant residues can be composted, as the Chinese do, but, the animals recover some of the energy, which otherwise will be entirely lost, in the composting process.

There are two types of livestock raising systems:

1. Sedentary
2. Transhumance

Animals in the first category are kept under a sedentary system and kept in and around a village throughout the year. All animals, excepting buffaloes, are put in daytime grazing, and, at night, crop residues and tree leaves are provided as fodder. Under the second category, animals move to different locations, depending on crop seasons. All animals during summer, when they are herded at high altitudes, mainly in alpine meadows, virtually stay on grazing.

In the high altitude mountain range, buffaloes are the preferred animals. Livestock size and composition are dictated by ecological and socioeconomic conditions. Cattle size is mainly decided by draught requirements. There is a misconception that people's reverential attitude towards cows (especially in India), is the factor contributing to livestock size. The size of buffaloes and goats have been increasing over the years as compared to that of cows. Cattle husbandry in Central Himalayas revolves around the requirement for agricultural draught and dung for manure. Small, yet sturdy, and sure-footed local animal breeds are preferred than improved or exotic ones, which do not thrive well at high altitudes. In the Upper reaches of the Himalayas sheep is the preferred type. The agro-pastoral society developed a lasting farming system, with a high degree of animal husbandry in animal breeding, crop production and land management.

Household Scenario

By convention, households do not come under the ambit of an agroecosystem. However, if one critically looks at a household, it becomes apparent that it is a part of it, because, agriculture is, indeed, a play of energy and capture of nutrients. Have we ever asked the critical question—"Where do most of the nutrients go to?". If we carefully analyze the situation, we would find that the nutrients, which are a pack of energy, go to sustain the members of a household. In fact, the human species is the biggest consumer of this pack of energy, known as nutrients. Every plant/crop, starting from its seed to all the other consumable vegetative parts, is packed with "nutrient energy", which is the sustenance for humankind and animals. The energy and nutrients which come from the soil to households are critical not just for a family's welfare, but also, for the productive performance of agriculture, on the whole. In fact, they are the basis of life- for humankind, animal kind, and, in fact, for other plants growing in the same soil. It is in this context that soil was called as "soul of infinite life" by Nair (2023a), during the World Soil Science Congress, held in Hamburg, The Federal Republic of Germany, in August, 1986.

A part of the energy transferred to the households, through consumed food, is expended into agriculture through human labor. Energy transferred through feeds (straws, fodder, weeds, and food grains) is expended again back in agriculture, by using draught animal power—by bullocks, mules and donkeys, in the mountain range/agriculture. This is what makes it so very unique.

It is important to note that in households, there are nutrients in kitchen waste, human waste, and animal waste, like poultry, pig waste. It is vital to recycle these nutrients back into soil resources, from which they came initially. To maintain sustainability, management of nutrient recycling from households to cultivated areas must be a subject of deep study/investigation. When this flow of nutrients is not intelligently managed, nutrients to cultivated land have to be brought from external sources, that is, the market. Excessive dependence on external inputs creates many ill effects on agriculture through sustainability—one of the most important lessons learnt from the after effect of highly soil extractive farming (Nair, 2023a). Hence, a household serves as an integral component of mountain agroecosystem. Additionally, the households serve as a foundation for household-based cropping system. This system is different from the run-of-the mill "kitchen gardening" or "backyard farming", mostly practiced in urban areas, where the little patch of land in the backyard is put vegetable cultivation or floriculture. This system cultivates vegetables, (olericulture) fruits (horticulture) or flowers (floriculture) for domestic consumption. This is very limited in scope. However, the current discussion is much larger in scope, since it aims to recycle nutrients emerging from households as an important source of nutrition for crops growing in the mountain agroecosystem. In the plains, where not-so-common kitchen garden or backyard farming is practiced, inputs are bought in the market, and brought to the cultivated fields to sustain cultivated crops. In the mountain agroecosystem, the source of nutrients is internally generated.

The Relevance of the Crop Wild Relatives (CWRs) in Mountain Agriculture

The Relevance of CWRs in the Curet Situation of Global Warming

Global warming is a reality mankind has to live with. Future food security will be dependent on a combination of the stresses, both biotic and abiotic, imposed by climate change, variability of weather within the growing season, development of cultivars more suited to different ambient conditions, and, the ability to develop effective adaptation strategies, which allow these cultivars to fully express their genetic potential under the changing climate conditions. Those plant species, which are very closely related to field crop, including their progenitors, having the potential to contribute beneficial traits for crop improvement, such as, resistance to an array of biotic and abiotic stresses, enriching the gene pool, leading ultimately to enhanced plant yield, thereby aiding humanity's relentless search for production of more food, to meet the ever growing needs of a burgeoning world population are called the "Crop Wild Relatives (CWRs)". In fact, CWRs are known to have tremendous potential to sustain and enhance global food security, thereby contributing enormously to humanity's well being. Therefore, their search, characterization and conservation in crop breeding programs assume great importance. Viewed against the recent upheavals in global climate change, the task becomes all the more important. Against the background of the disastrous after effects, especially alarming environmental hazards, much of it soil-related, of the highly soil extractive farming, euphemistically known as the "green revolution", of the 1960s (Nair, 2019a), the task assumes much cruciality. It is in this context that the current brief description of the relevance of CWSRs to global biodiversity, inasmuch as mountain agriculture is concerned, this discussion in this book is initiated.

The relevance of "Crop Wild Relatives" has been extensively discussed, with reference to their utility in combating global warming (Nair, 2019b). However, this chapter discusses their list, which could be of relevance in mountain agriculture. Table 2.3, catalogues CWRs of economic importance in the Himalayan region.

The appropriate environmental conditions have been provided for several varieties of cereals, pulses, fruits, oilseeds, spices, tuberous vegetables and sugar-yielding plants and their close relatives that number about 155 in the Himalayan Mountain region (Arora and Nayar, 1984, refer Table 2.3). The global crop plant genetic resources can be, broadly, assigned to the specific centers of diversity identified by Vavilov (1951), based on the varietal diversity, homologous variation, endemism, dominant frequency of alleles and resistance to pests and diseases. The Indian sub continent can be divided into the following eight sub centers:

1. Western Himalayas
2. Eastern Himalayas
3. North-Eastern Region
4. Gangetic Plains
5. Indus Plains
6. Western Ghats
7. Eastern Ghats
8. Andaman and Nicobar Islands in The Indian ocean

In fact there are about 399 species of Crop Wild Relatives (CWRs) of crop plants, which belong to 90 genera in the western, eastern and north eastern Himalayas, which show the enormous potential of the CWRs to tap their utility to enhance the productivity and/or quality of the current varieties of crop plant being cultivated, some of which, especially the high yielding varieties (HYV) or high responsive

Table 2.3 CWRs of economic importance in the Himalayan region

A. North and North-West Himalaya	
Crops	Botanical description
Cereals and millets	*Aegilops tauschii, Avena barbeata, A. fatua var. fatua, A. ludoviciana, Digitaria sanguinalis, Elymus dalutricus, E. dasystachys, E. nutans, E. distans, E. orientale, Hordeum galucum, H. spontaneum, H. turkestanicum, Pennisetum orientale*
Legumes	*Cicer microphyllum, Lathyrus aphaca, Moghania vestita, Mucuna capitata, Trigonella emodi, Viga capensis, V. radiata var. sublobata, V. umbellata*
Fruits	*Elaegans hortensis, Ficus palmate, Fragaria indica, Morus* spp. *Pruns acuminata, P. cerasoides, P. cornuta, P. napalensis, P. prostrata, P. tomentasa, Pyrus buccata P. communis, P. kumaoni, P. pashia, Ribes glaciale, R. nigrum, Rubus ellipticus, R. fruiticosus, R. lanatus, R. lasiocarpus, R. molluccanus, Zizyphus vulgarris*
Vegetables	*Abelmoschus manihot (tetrahyllus forms), Cucumis hardwickii, C. callous, Luffa echinata, L. graveolens, Solanum incanum, S. indicum, Trichosanthes multiloba, T. himalensis*
Oilseeds	*Lapidium capitatum, L. latifolium, L. draba, L. Ruderale*
Fibres	*Linum perenne*
Spices and condiments	*Allium rubellum, A. schoenoprasum, A. tuberosum, Carum bulbocastinum*
Miscellaneous crops	*Saccharum filifolium, Miscanthus nepalensis*
B. Eastern and North-Eastern Himalayas	
Cereals and millets	*Digiaria cruciata, Hordeum agricrithon*
Legumes	*Atylosia barbata, A. scarabaeoides, A. villosa, Canavalia ensiformis, Mucuna bractearata, Vigna umbellate, V. radiata var. sublobata, V. pilosa*
Fruits	*Fragaria indica, Morus* spp., *Myrica esculenta, Prunus acuminata, P. cornuta, P. jenkinsii, P. nepalensis, Pyrus pashia, Ribes glaciale, Rubus lineatus, R. ellipticus, R. lasiocarpus, R. molluccanus, R. reticulates, Citrus assamensis, C. ichangensis, C. indica, D. hookeriana, Eriobotya augustifolia, Mangifera sylvatica, Musa balbisiana complex, M. manii, M. nagensium, M. sikkimensis, M. supreba, M. velutina*
Vegetables	*Abelmoschus manihot (pungens forms), Luffa graveolens, Neoluffa sikkimensis, Cucumis hystrix, C. callosus, Momordica dioica, M. cochinchinesis, M. macophyllata, M. subangulata, Trichosanthes cucumerina, T. dioica, T. dioaelosperma, T. khasiana, T. ovata, T. trumcata, Solanum indicum and tubertypes, Allocasia macrohiza, Amorphophallus bulbifer, Colocasia esculenta, Dioscorea alata, Moghania vestica, Vigna capensis*
Oilseeds	*Brassica trilocularis types*
Fibers	*Gossypium arboretum* (primitive types)

(continued)

Table 2.3 (continued)

A. North and North-West Himalaya	
Crops	Botanical description
Spies and condiments	*Allium tuberosum, A. sublobatum, Curcuma zedoaria, Alpinia galanga, A. speciosa, Amomum aromaticum, Curcuma amoda, Piper peepuloides*
Miscellaneous CWRs	*Saccharum langisetosum, S. sikkimensis, S. ravennae, Erianchus* spp., *Miscanthus nudipus, M. nepalensis, M. tylorii, Naranga fallax, Camellia* spp.

Source Arora and Nayar (1984)

varieties (HRV), of wheat and rice, of the green revolution, which have fallen prey to the attack of both insect pests and diseases, in addition to the vicissitudes of global warming (Nair, 2019b, 2019c).

References

Arora, R. K., & Nayar, E. R. (1984). *Wild relatives of crop plants in India.* National Bureau of Plant Genetic Resources (NBPGR).

Jodha, N. S. (1990). *Mountain agriculture: Search for sustainability. Mounting farming systems discussion paper no. 2.* ICIMOD.

Jodha, N. S., Banskota, M., & Partap, T. (1992). *Sustainable mountain agriculture.* Oxford & IBH Co.

Knorr and WWF. (2019). Future 50 foods for healthier people and a healthier planet. 60pp. https://www.knorr.com/content/dam/unilever/knorr_world/global/online_comms_/knorr_future_50_report_online_final_version-1539191.pdf.

Nair, K. P. P. (2019a). *Intelligent soil management for sustainable agriculture: The role of nutrient buffer power concept*. Springer Nature, Switzerland, AG.

Nair, K. P. P. (2019b). *Combating global warming—The role of crop wild relatives for food security.* Springer Nature.

Nair, K. P. P. (2019c). Utilizing crop wild relatives to combat global warming. *Advances in Agronomy, 153*, 175–258.

Nair, K. P. P. (2023a). *The living soil: A lifetime journey in understanding it for human sustenance.* SpringerBriefs in Environmental Science.

Nair, K. P. P. (2023b). Extractive farming or bio farming. In *A better choice for 21st century*. Springer Nature.

Singh, V. (2005). *Agrobiodiversity, sustainability and food security in the Himalayan mountains: An Uttaranchal perspective.* Gorakpur: Gorakpur Environmental Action Groip. 50pp.

Singh, V. (2018). Food security amidst the climate change scenario: Perspectives, issues and opportunities in mountain agriculture. *International Journal of Food Science and Nutrition, 6*(3), 555686. https://doi.org/10.19080/NFSIJ (218.06.555686).

Singh, V. (2022). *Three crops rule the world: What it means for the planet's wildlife*? State of India's Environment.

Singh, V., Shiva, V., & Bhatt, V. K. (2014). *Agroecology: Principles and operationalisation for sustainable mountain agriculture* (64p+viii).

Vavilov, N. I. (1951). The origin, variation, immunity and breeding of cultivated plants. *Chronica Botanica, 13*, 1–364.

Chapter 3
Biodiversity of the Pedosphere

The immense biodiversity that Nature possesses is not immediately visible to our naked eye. The human naked eye is incapable of investigating the vastness, depth and beauty of Nature's biodiversity. A vast amount of Nature's biodiversity flourishes in the ecosystems not directly accessible to us and the organisms in the sizes that our naked eyes cannot record. In fact, for example, the soil biodiversity is so very vast and fascinating than what we have so far imagined. Nair (2023a) in his most recent book "The Living Soil" has coined the phrase "Soul of Infinite Life" for soil.

The biodiversity in the soil, together with the physical environment, is what constitutes the pedosphere. There is a biodiversity in the soil, which is underground, as different from the biodiversity above the soil surface, which has been discussed so far in this book. In 1992 the historical Earth Summit was held in Rio de Janeiro, Brazil, where the Convention on Biodiversity (CBD) underlined the significance of biodiversity conservation for various tangible and intangible benefits, as well as the key factors for sustainability of all forms of Life on Planet Earth. Very disappointingly, during this dialogue, only scanty attention was paid to the conservation of the subject of soil biodiversity, to realize the full potential of the soil resources to maintain the sustainability of the terrestrial ecosystems. However, with the beginning of the 21st century, due emphasis to soil community conservation was given in international environmental policies, specifically, in the European Union Soil Thematic Strategy (2006), the Biodiversity Action Plan for Agriculture (EU 2001), and, the Kyiv Resolution on Biodiversity (EU/ECU 2003) (Menta, 2012).

What is the Pedosphere?

The outermost layer of Planet Earth composed of soil, which is subject to soil formation processes, an integral part of the biosphere, lying on the interface of atmosphere, hydrosphere, and lithosphere, and a lot more complex in its physico-chemical and biological characteristics, and functioning, which is quite distinguishable from that

K. P. Nair, *Biodiversity in Agriculture*,
https://doi.org/10.1007/978-3-031-44252-0_3

of the above-soil surface terrestrial and aquatic ecosystems, is the pedosphere. In the pedosphere, there are an array of organisms organized in trophic levels of auto and heterotrophs, which make up a variety of communities of independent organisms in a variety of soil ecosystems. What is the physical/environmental parameters of the pedosphere:

1. Air (Gases)
2. Water
3. Moisture
4. Organic and inorganic substances

The inorganic substances include a variety of minerals—metallic and non-metallic, including all the macro or major nutrients, such as nitrogen phosphorus and potassium, secondary nutrients, such as calcium, magnesium, sulfur and micro nutrients, such as, zinc, copper iron, manganese etc. The organic component consists mainly of soil organic matter (SOM). The SOM consists of the following:

1. Fresh plant residue—stubble after harvest
2. Small living organisms decomposing the fresh plant residue
3. Stable organic matter—Humus

Organic substances include dead remains of plant and animal substances and compounds such as carbohydrates, proteins, fats, oils, nucleic acids etc., along with humus, the stable organic matter—which has attained complete decomposition.

There is a rich diversity of living species, their genotypes, and communities, which make up the biotic component of the soil. These are dominated by microorganisms, and the pedosphere also embraces fauna and flora belonging to higher living forms. The organisms belong to the five kingdoms of life—monera, protista, fungi, plants and animals. The life in the pedosphere determines the above soil surface life. In fact, the pedosphere is so very critical and important to the entire biosphere.

What is Pedodiversity?

On planet Earth, there is no other domain, as great as pedodiversity, in the entire gamut of biodiversity. It constitutes one of the unique characteristics of the soil ecosystem, or pedoecosystem, to be more specific. The pedodiversity is a very huge part of the entire biodiversity. Soils of the planet Earth vary according to their geographical positioning, the climate parameters, physical structures, chemical and biological features. The vegetation cover influences the soil environment. As human activities determine, to a large, extent, the vegetative cover of the soil resources, it can be said that anthropogenic factors influence greatly what the soils will be in course of time. Extended, these activities can completely destroy the soil resources, as has happened, as a consequence of the ill-thought and ill-fated "green revolution" (Nair, 2019a), clearly enunciated by Nair in his most recent book "The Living Soil" (2023a).

The planet Earth's soils have been broadly classified into the following categories:

1. Sandy
2. Silty
3. Clayey
4. Loamy
5. Chalky

The above are made up of the three major types:

1. Sandy
2. Clayey
3. Silty (Loamy)

Of the above three mentioned, sandy soils have the particle-size fractions, classified between 0–20 μm or 0–50 μm for the smallest, clay fractions, according to the sol classification followed (European or American). Within a soil ecosystem, the soil particle sizes determine the space sizes and hence water drainage/water and nutrient holding quality of the soils, and, as a consequence, the quality of the soil environment necessary for soil biodiversity and the status of plant growth. Among the three soil types, loamy type is regarded and "perfect" with appropriate proportion of the three major soil types—sand (40%), silt (40%) and clay (20%). And this type of soil contains a high proportion of soil carbon indicating high soil fertility. This type of soil is also considered to be the most ideal for crop production/cultivation.

The Invisible Biodiversity

The pedosphere is richest to an invisible biodiversity, because, soil contains the following:

1. In a single square meter of soil, more than 1000 species of invertebrates can be found
2. At least during some stage of their life cycle, planet Earth is the abode for most of the terrestrial insect population
3. A gram of soil is host to thousands of bacteria species
4. In a healthy soil, one can find thousands of the species of bacteria, fungi and actinomycetes, tens of nematode species, 50–100 species of insects, 20–30 species of mites, several species of earthworms, and several species of invertebrate animals
5. Soil contains the organisms with the largest area, for example, a single colony of honey fungus (*Armillaria ostoyae*), may cover approximately 9.0 km^2

The "invisible world" beneath our feet, which the soil encompasses, blossoms with the biodiversity, which measures more than the that of the biodiversity of the "visible world" outside, in all aspects, which humans cannot fathom. These factors encompass the spread of the various species, their density, their genotypes, distribution, and also, biomass. But, the "invisible" biodiversity depends on the ecological state of the soil ecosystems. Living soil destroyed by desertification and/or damaged by the extractive farming (as explained in the most recent book by Nair (2023a), is likely to harbor only a small fraction of the entire biodiversity. It is most fascinating to observe that the biodiversity of the "visible world" is rooted in the biodiversity of the "invisible world". Simply put, in the state of biodiversity of the terrestrial ecosystems one can witness, with one's own eyes, the immense biodiversity of the soil ecosystems, in other words, the pedosphere. In photosynthesis, chemosynthesis, fixation of nitrogen, sedimentary cycles etc., the criticality of the pedosphere biodiversity becomes apparent. Chemosynthesis, chiefly controlled by soil bacteria, and further, photosynthesis by cyanobacteria, and, algae, influence the primary productivity of a soil ecosystem, and, both the autotrophic mechanisms, namely, chemosynthesis and photosynthesis, become the basis of secondary productivity. A key role is played by photosynthesis in the above-soil terrestrial ecosystems, infusing carbon in the pedosphere, which also feeds a variety of detritivores, (a type of heterotrophs which feed on dead and decaying organic matter to meet their nutrition and energy requirements) in addition to enriching soil fertility. Most of the soil biodiversity of heterotrophs is supported by the organic matter emanating from photosynthesis, in the terrestrial ecosystems.

A few microorganisms in the soil ecosystem, as for example, blue-green algae or cyanobacteria, are involved in a dual role, that is, photosynthesis and nitrogen fixation. The photosynthesis performed by these organisms also contributes to the organic matter enrichment of the soil mass, thereby, vastly enhancing the fertility status of the entire ecosystem, via, soil resources.

The soil microflora, which comprises of the bacteria, actinomycetes, fungi, cyanobacteria and algae, represents a very high level of soil biodiversity, within the larger umbrella of general biodiversity. These organisms are very vibrant, and more varied than the soil fauna, which includes protozoa, nematodes, collembola or springtails (which are small arthropods, also considered to be the most abundant macroscopic organism on planet Earth), and acarids, which tend to move about upon exhaustion of their foods. The soil organisms can be classified into the following three groups:

1. Micro flora and micro fauna
2. Mesofauna and meiofauna
3. Macro fauna

The flora of meso and macro sizes inhabit land surfaces, with roots in the soil. And hence, are not considered as soil-bound organisms.

The Soil Flora

The soil flora is an integral part of the whole soil organic matter (SOM), the reservoir of soil biodynamics. In fact, the pedosphere is the important repository of the soil microflora, which, in entirety forms the biotic component of the soil.

The Soil Microflora

Among the soil microflora, the dominant component is the bacteria, and, this makes up around 90% of the entire soil microflora population. Around 9% of the microflora population of the soil is made up of Actinomycetes, and, the last fraction, of about 1% is constituted of fungi and algae. All of the biodynamic processes in the soil are influenced by these microflora. But, the bacterial population play the pivotal role in these biodynamic processes. In fact, the soil health is greatly influenced by the microflora, and, one can imagine to what extent, the microbial, in fact, bacterial, population play in these biodynamic processes. Just to provide an example: the entire nitrogen mineralization process can be ruined/or can be at stake, if the activity of *Nitrosomonas* bacteria is not followed smoothly by the activity of *Nitrobacter* bacteria. This is, because, the former converts the ammonium ion to nitrite ion and the latter converts nitrite ion to nitrate ion. In fact, if this sequential bacterial activity does not function smoothly, it can lead to nitrite poisoning of any organism, for example plant, that grows in the soil. Hence, microflora, in essence, is the key biotic factor which controls the pedosphere health. In fact, it is the microflora that controls the flow of energy in the pedosphere.

The soil bacterial population flourish when soils, for instance, those whose pH revolves around 6.9–7.1, are supplied with huge amount soil organic matter. Among the microflora, bacteria are both autotrophs and heterotrophs. The autotrophic bacteria utilize inorganic molecules, as their source of energy and atmospheric carbon dioxide as a source of carbon, for their nourishment, through the process known as chemosynthesis. Hence, the chemosynthesis bacteria are also known as chemoautotrophs. For energy, the chemoautotrophic bacteria depend on a variety of inorganic substances, and, are classified as the following:

1. Nitrifying bacteria
2. Sulfur bacteria
3. Iron bacteria
4. Manganese bacteria
5. Methane bacteria
6. Carbon monoxide bacteria

Algae and blue-green algae, and, the cyanobacteria are photosynthetic organisms. These are also classified as photoautotrophs. The most of the soil bacteria are however, heterotrophs. The heterotrophic bacteria depend on organic matter and carbon (which originate in SOM,—soil organic matter) as a source of energy, and are mainly concerned with the decomposition processes in the soil leading to mineralization of plant nutrients, which are essential for the sustenance of plants. Most of the essential minerals needed for plant nutrition are released during the decomposition process mediated by soil bacteria.

Soil biodiversity is specifically composed of the following:

1. Nitrogen- fixing bacteria
2. Actinomycetes
3. Cyanobacteria (blue-green algae)

All of the above three are involved in the process of biological nitrogen fixation (BNF). In this connection, it must be said that the atmosphere contains about 78–80% of elemental nitrogen, but, this cannot be utilized by plants, like wheat, rice and maize, except the legume species, like red gram, and vegetable species of the legume category, like the vigna species, *Vigna ungiculata* (L.). These bacteria include the following:

1. Free living bacteria and symbiotic bacteria are the following:
   A. Aerobic bacteria and anaerobic bacteria: *Clostridium, Azotobacter* (involved in atmospheric nitrogen fixation)
   B. Blue green algae or Cyanobacteria (photosynthetic, autotrophs inhabiting waterlogged areas and wet soil, such as Anabaena, Nostoc, Scytonema, Algae)
2. Symbiotic bacteria with several fungi and nodule-forming bacteria (such as the *Rhizobium*), involved in atmospheric nitrogen fixation, and these establish a mycorrhizal association—ecotropic and endotropic—with the roots of certain plants, such as, Pinus, Casuarina and nodule forming bacteria, e.g. Rhizobium species:

It is in acidic (less than pH 6), soils that fungi inhabit. The fungi are parasitic in nature and for their nourishment, they utilize saprophytes, and, symbiotics. The roots of plants are infected by parasitic fungi, causing several serious plant diseases, such as root rot, wilts, blights, rusts and smuts. Some of the wilt-forming parasitic fungi release chemical substances, which are extremely harmful to the infected plants, which ultimately die. For instance, *Fusarium oxysporum* f. sp. *lini*, a fungus responsible for producing wilt in the flax plant, produces hydrogen cyanide (HCN), which is a poison, and another *Fusarium udum*, causes wilt disease of pigeon pea (*Cajanus cajan*) in the tropical regions, releases fusaric acid in the roots of the host plant they infest. Certain parasitic fungi also secrete substances that function as growth stimulants, which indeed, is a beneficial effect. *Fusarium* spp. are known to produce a hormone, gibberlic acid ($C_{19}H_{22}O_6$), also called as gibberlins, A3, GA, and GA3, which benefits the host plants by enhancing their growth.

The saprophytic fungi derive its energy from the carbon of the organic matter in the soil. During the decomposition process, they break down cellulose, lignin, starch, sugars, and proteins etc., into their constituent components. Many fungal species inhabiting soil have demonstrated their antibiotic properties. These antibiotic compounds are of great societal value. Some of the fungal species isolated from the soil have demonstrated antimicrobial properties, of the which the following are important:

1. *Absidia corymbifera*
2. *Aspergillus flavus*
3. *Alernaria alternata*
4. *Aspergillus fumigants*
5. *Aspergillus niger*

6. *Cladosporium herbarum*
7. *Cuvularia lunata*
8. *Penicillium* sp.
9. *Rhizopus stolonifer*
10. *Trichoderma viride*

Number 10 is vastly used in the control of "Quick Wilt" disease of Black pepper (*Piper nigrum*) caused by the *Phytophthora* fungus.

The Soil Macroflora

Macro flora is part of the ecosystem above the soil surface, though a significant proportion of it lies under the soil surface. The following five ecological classes encompass the soil macroflora:

1. Halohytes: These are plants occurring in saline soils, e.g. *Salsola foetida*
2. Psammophytes: These are plants growing on sandy soils, e.g. *Agriophyllum squarrosum*
3. Lithophytes: They grow on rock surfaces, e.g. *Dendrobium*
4. Chasmophytes: Plants growing in rock crevices, e.g. *Asarina procumbens*
5. Oxylophytes: Plants growing in acid soils, e.g. *Rumex* sp.

The Soil Fauna

Soil fauna is one of the most interesting components of the soil ecosystem, which performs several functions. These soil fauna are important indicators of the quality of soil—its fertility status, its physical status and its biological status. They are as follows.

The Soil Microfauna

In this category, animals, whose size varies from 20 to 200 μm are placed, while some researchers put the upper limit to 100 μm. Some of these animals are the following:

1. Protozoa
2. Small-sized mites
3. Nematodes
4. Rotifers
5. Tardigrades (eight-legged segmented micro-animals), and, copepod crustaceans.

The above mentioned organisms live up to different depths in the soil. For instance, protozoans, like ameba, ciliates, and, zoomastigina flagellates, live near the soil surface, while the testate forms, such as, *Thecamoeba*, *Euglypha*, and *Diffugia*, are scattered vertically. Nematodes are found in abundance in soils. For instance, nematodes, such as, *Criconemoides* and *Aphelenchoides* prevail, as plentifully as 1–3 million per square meter in soils with raw humus, while grassland soils can contain up to 20 million per square meter.

The Soil Mesofauna

The animals included in this category measure 200 μm to 1 cm. According to some investigators, it is 0.1–2 mm size. Micro-arthropods Acarina (mites) and Collembola (springtails) make the important members of this group which includes the larger nematodes, rotifers, and, tardigrades, along with the majority of isopods, Arachnida, Chelognathi, Opiliones, Enchytraeidae insect larvae and small millipedes, isopods and mollusks. The soil mesofauna density and populations vary depending on the soil composition, especially, the ratio between soil organic matter and soil moisture content. The soil mesofauna population and density are also influenced by the specific soil management conditions.

The Soil Macrofauna

Those animals inhabiting soil, which are sized more than 1 cm fall under this category. The body size of the soil inhabiting animals designated as macrofauna is between 1–2 and 20–30 mm (Gongalsky, 2021). In some ecosystems, the macrofauna may account for most of the total soil animal biomass which substantially contributes to the functioning of soil food-web (Gongalsky, 2021). The major types of soil animals include those belonging to Lumbricidae (a family of earthworms). Mollusca (the second largest phylum of invertebrates after the Arthropoda), and also large-sized chilopods, arachnids, insects and fossorial vertebrates (amphibians, reptiles, birds and mammals), are the other animal families of soil macrofauna.

Among the soil macrofauna, the earthworms occupy a place of great importance. They decompose organic matter, break down litter fragments, mixing up organic matter with the soil. The practice of vermicomposting is very important, where earthworms play a crucial role. The common Indian Annelid species of earthworm, dwelling in the soil are the following:

1. *Pheretima* (found mainly in New Guinea and parts of south Asia)
2. *Megascolex* found in soils of Madagascar, Australia and New Zealand, North America and South East Asia
3. *Octohaetus*—native to Australia, New Zealand and Malaysia

4. *Drawida* commonly found in India and China
5. *Monilgaster* generally restricted to Western Ghats in India
6. Numerical dominance of the earthworms is found in alkaline soils with high organic matter content and adequate amounts of soil moisture. In acidic soils, earthworms are sparsely found. The above classification is by Narayanan et al. (2022)

The centipedes species *Scolopendra* and the stone centipedes *Lithobius* among chilopods and the European mole cricket widely distributed in Europe, *Gryllotalpa* and an omnivorous in *Forticula, Forrticula auricularia*, the common earwig or European earwig, among the insect population, are well adapted to fossorial life in soil habitats. For land snails, soils rich in lime are more suitable.

The Soil Megafauna

In both soils, below the land surface, and, terrestrial surface, soil megafauna can thrive. These are soil vertebrates. In the soil, the vertebrates usually dwell in fossorial or in a "burrowing" lifestyle, usually, digging holes and shifting soil and living in such holes. The species commonly known as "rain frogs" or short-headed frogs, are soil vertebrates, among Amphibians, which includes *Breviceps* and *Ichthyopis* (a genus of caecilians—which are the limbless amphibians, which are sometimes called the Asian caecilians), and, among reptiles, which includes Sphenodon, which is the last survivor of the reptilian order *Rhynchocephalia*, believed to have evolved in the early Mesozoic era, *Uromastyx*, a limbless lizard (a genus of the African and Asian agamid lizard), limbless lizards and snakes. Among the Avian species, it is the burrowing owl. And among the common mammals, we have the kangaroo-rats, ground squirrels, otter, moles, rodents, badgers, hedgehogs (a spiny mammal) etc.

What is the "Erosion of Soil Biodiversity"?

The soil type, soil composition, climate, topography, ambient temperature etc., determine the soil biodiversity. Soil biodiversity is very adversely affected due to many of the natural factors, such as, soil degradation, the biggest environmental hazard of the chemically-centric green revolution, soil erosion, etc., eventually leading to soil desertification. It is the anthropogenic factors (which are man-made) which are the primary cause for the loss of soil biodiversity. The offshoot of this anthropogenic factors is the land use changes, a major cause of erosion of soil biodiversity. Transformation of forested land, rangelands and grasslands into cultivated lands, is the most illustrious example of the erosion of soil biodiversity. This leads, ultimately, to a crisis in the soil ecosystem.

As an example of the severe impact of soil degradation, of India's 328.73 million hectares of geographical area, more than 120.40 million hectares have now degraded soils, mainly in Punjab State, the "cradle" of Indian green revolution. As per one important estimate, a huge chunk of agricultural land in the world—approximately 40%—is now degraded (Sample, 2007). The Special IPCC Report (IPCC, 2019) further makes a crucial point about land degradation, as follows:

1. Close to a quarter of the planet Earth's ice-free land area is subject to human-induced degradation
2. Soil erosion rate, compared to soil formation rate, from agricultural fields is currently estimated to be 10–20 (where there is no tillage) to more than 100 times (where there is conventional tillage)

A land area covered with dense perennial vegetation, for instance, a natural forest, builds up to a soil ecosystem rich in biodiversity. When an area is deforested and converted into cultivated lands, the soil erosion process is triggered. Intensive tillage, frequent irrigation, continuous monoculture, unbridled use of synthetic chemical fertilizers, pesticides and herbicides, all lead to soil erosion. Mismanagement of the soi in a specific area further escalates the processes that on a long-term basis end up as an unproductive desert. A natural forest evolved up to its ecological climax through succession blossoms with highest degree of biodiversity. Deforestation followed by unabated land degradation eventually leads to a state of disclimax via the process of desertification which represents minimum to zero biodiversity.

How to Conserve Soil Biodiversity?

Soil is home, both to terrestrial and sub terrestrial life, and, it is very vital to preserve it, in as healthy a state as possible, for the preservation and perpetuation of all life—human, animal and plant. When soil is destroyed, either by man's ill thought intervention, or through nature's fury, life is at stake for all. Vegetation cover on the soil surface, especially the natural forests with trees, as the dominant members of the forest community, will be a major determinant of biodiversity conservation. The soil surface biodiversity (the forest biodiversity) influences soil biodiversity and vice versa. Preservation and enhancement of existing forests, afforestation on a massive scale in the deforested area, coupled with the task of reclaiming degraded land, through recourse to various ecological and soil ameliorative processes, is the priority task to ensure pedosphere diversity.

Current agricultural practices, centered on an excessive soil extractive scale, is bound to hugely impact, negatively, on pedosphere biodiversity (Nair, 2023a). Ecologically sound soil fertility management techniques, cropping systems involving a synergistic combination of deep and shallow-rooted crops, greater emphasis on agroforestry, and, no till farming, and, reestablishment of forestry-cropping relationships, are some of the approaches that will greatly ameliorate the current unhealthy state of soil environment and pedosphere biodiversity. Time is running out on this crucial and critical task. Life on planet Earth is at stake. Intelligent soil management is the need of the hour (Nair, 2019a).

In summary, it can be said that the pedosphere has tis specific environment which supports and sustains an amazing string of organisms, organized in trophic levels of auto and heterotrophs, making up a variety of communities interdependent organisms in a multiple of soil ecosystems.

Population and density of soil mesofauna vary as per the soil composition, especially the proportions of organic matter and soil moisture contents.

References

Gongalsky, K. B. (2021). Soil macrofauna: Study problems and perspectives. *Soil Biology and Biochemistry, 159*, 108281. https://doi.org/10.1016/j.soilbio.2021.108281

IPCC. (2019). Summary for policymakers. In P. R. Shukla, J. Skea, E. Calvo Buendia, V. Masson-Delmotte, O. Portner, D. C. Roberts, P. Zhai, R. Slade, S. Connors, R. van Diemen, M. Ferrat, E. Haughey, S. Luz, N. Neogi, M. Pathak, J. Petzold, J. Portugal Pereira, P. Vyas, E. Huntley, K. Kissick, M. Belkacemi, & J. Malley (Eds.), *Climate change and land: An IPCC special report on climate change, desertification, land degradation, sustainable land management, food security, and greenhouse gas fluxes in terrestrial ecosystems.* IPCC.

Menta, C. (2012). Soil fauna diversity—Function, soil degradation, biological indices, soil restoration. In G. A. Lammed (Ed.), *Biodiversity conservation and utilization in a diverse world.* IntechOpen. https://doi.org/10.5772/51091

Nair, K. P. P. (2019a). *Intelligent soil management for sustainable agriculture: The nutient buffer power concept.* Springer Nature, Switzerland, AG.

Nair, K. P. P. (2023a). *The living soil: A lifetime journey in understanding it for human sustenance.* SpringerBriefs in Environmental Science.

Narayanan, S. P., Anuja, R., Thomas, A. P., & Paliwal, R. (2022). A new species of Moniligaster Perrier, 1872 (Annelida, Moniligastridae) from India, with status revision of M. deshayesi minor Michaelsen, 1913. *Opuscula Zoologica, 53*(1), 31–50. https://doi.org/10.18348/opzool.2022.1.31

Sample, I. (2007). *Global food crisis looms as climate change and population growth strip fertile land.* The Guardian, August 31, 2007.

Chapter 4
Chemosynthesis-Based Community Bio Diversity

In the biosphere, nutrition, is at the center—be it for humankind, animalkind or plantkind. This is the basis of life all the three. The following are two modes of nutrition for all species:

1. Self nourishment
2. Cross nourishment

Autotrophs employ self nourishment. Heterotrophs employ cross nourishment. For instance, a saprophytic fungus, feeds on dead and decaying life—be it a plant cell or an animal cell, however small it is. The autotrophs nourish themselves by utilizing inorganic molecules and the energy available in the nature, while heterotrophs derive the nutrients that the autotrophs manufacture for themselves. This means, heterotrophs derive nutrients from external organic molecules. These organic molecules are derived either from the autotrophs directly or from the consumers dependent on autotrophs.

Light is the source of energy for autotrophs. Heterotrophs harness energy bound to organic molecules. In substance, nourishment of the heterotrophs originate in the energy of organic molecules derived from other autotrophs.

Which Are the Two Types of Living Communities?

The following are the two communities based on their nutrition mode:

1. Those communities which are photosynthesis-based
2. Those communities which are chemosynthesis-based

The first type comprises all the terrestrial populations, most of the organisms in the soil and aquatic ecosystems. The second type are those in the soil, as well as, cold seeps, sunken woods, vertebrate falls—which are the carcasses of whales—and those on ocean floor around hydrothermal vents.

K. P. Nair, *Biodiversity in Agriculture*,
https://doi.org/10.1007/978-3-031-44252-0_4

The Communities Based on Photosynthesis

The inorganic ions, CO_2 and H_2O which are synthesized together in the presence of light and chlorophyll is photosysnthesis. In this reaction, light forms the energy source. Light, a thermal energy, is converted into a biochemical energy. In fact, light is living energy or energy of life, and this is bonded with sugars in living organisms. In substance, photosynthesis is a process in which light transforms itself into energy binding together CO_2 and H_2O, within chloroplasts found in green plants. Algae, and cyanobacteria, and, released oxygen molecules. Hence, only chlorophyllous organisms can perform photosynthesis and subsist as autotrophs.

All biomolecules are synthesized through the mechanism of photosynthesis, which is a capture of the solar energy by chlorophyllous organisms, which help in the maintenance, growth and reproduction of the organism. Additionally, this energy also supports a vast number of heterotrophic organisms, in communities ranging from herbivores to carnivores. The producers of the energy are the autotrophs. The energy flow from autotrophs to other trophic levels constituted by consumers, such as, herbivores, primary carnivores, secondary carnivores, and topmost carnivores, follow the 10% Law elaborated by Lindeman (1942). This process becomes the basis for the nourishment of a number of organisms. A substantial proportion of the organic matter produced by living photosynthesizers, and also from dead organisms—which are both producers and consumers—forms the detritus which also supports a vast number of detritus consumers in the soil substrate and also the sediment of aquatic ecosystems. Hence, photo synthesis is the foundation for the formation of life on planet Earth. This applies to both terrestrial soil substrate and aquatic ecosystems.

On planet Earth, among the communities of ecosystems, photosynthesis ensures an enormous amount of biodiversity at Alpha, Beta, and, Gamma levels. Each of the food chain, and each of the trophic level—the grazing, also the detritus—in the photosynthetic-dependent communities in all ecosystems represent an enormous amount of biodiversity. Hence, a greater part of Nature's life, inarguably, is photosynthesis-based biodiversity. Both the humankind and animal kind are exclusively dependent on the mechanism of photosynthesis for formation or synthesis of life and its further sustenance.

The Communities Based on Chemosysnthesis

The second category of organisms are those which are chemosynthesis-based. Microorganisms, which are specialized, belong to this category. And, the bacteria and archaea are the most illustrious examples.

It was only in the latter part of the seventies that the discovery of chemosynthesis-based communities was established. In fact, in the area of biology, this discovery was, indeed, intriguing. In terms of evolution, the following are the three locations/ecosystems where they evolved:

1. Hydrothermal vents
2. Cold seeps
3. Hot springs on land

The whale fall on the ocean floor also becomes a habitat for the chemosynthesis-based community. Sunken wood and soil communities are also, partially, chemosynthesis-dependent communities. The earliest known chemosynthetic association containing metazoan animals has been described by Kaim (2010) in the Ural Mountains.

Complex organic compounds are synthesized by chemosynthetic organisms, utilizing inorganic substances as a source of energy and CO_2 or CH_4, as sources of carbon. These autotrophs are also called chemoautotrophs in scientific literature. It is very important to note that chemosynthesis occurs in the absence of light, unlike photosynthesis. Chemoautotrophs inhabit places/ecosystems away from the reach of solar radiation. Fixing inorganic carbon through the oxidation of chemical compounds, the chemoautotrophs are phylogenetically diverse, representing taxa of biochemical interest, for instance, gammaproteobacteria, epsilon proteobacteria, aquificaeles, neutrophilic iron-oxidizing bacteria and methanogenic archaea (Singh, 2020).

It was the German scientist Wilhelm Pfeffer who coined the term chemosynthesis in 1897. Even before the discovery of Wilhelm Pfeffer, some other scientists, such as, the Russian scientist Sergei Nikolaevich Vinogradskii had shown that some other microorganisms are capable of living exclusively on inorganic substances. He referred to this process as "anoroxydant" in 1890, the term subsequently, rechristened as chemosynthesis. More recently, the term "chemolithoautotrophy" has been used to describe the process.

A number of chemosynthetic organisms prevail in Nature based on the chemical substances which they oxidize to use as sources of energy. The inorganic substances from which the chemoautotrophic microorganisms derive energy are the following:

1. H_2
2. CO
3. Fe^{2+}
4. NH_4^+
5. H_2S
6. S

Nearly, the entire biological nitrogen fixation process occurs due to the role of some specific chemosynthetic macroorganisms, which independently thrive in soils, water, (ponds), and, synergistically on the roots of a number of plants, especially, which belong to the family Fabaceae or Leguminosae. The nitrifying bacteria are of much critical ecological importance, as much of the ecosystem structure owes to the specific role of these bacteria in the nitrogen fixation process. Nitrogen incorporated into amino acids and then into protein is a basic structural molecule. These microorganism oxidize NH_4^+ to NO_2^- and subsequently to NO_3^- and derive energy for the synthesis of complex biomolecules, as per the following reaction:

$$2NH_4^+ + 3O_2^- \rightarrow 2NO_2^- + 4H^+ + 2H_2O + \text{Energy}$$

Methane, oil, and other hydrocarbon fluxes are observed in cold seep areas around earth's oceans, which make their way into the sediment-water interface from the bottom via relatively focused conduits, such as, faults (Arvidson et al., 2004). These areas are known as "hot spots" for chemosysnthesis, with methane and its decomposition by-products, such as H_2S (hydrogen sulfide, Milucka et al., 2012). The chemosynthesis which goes on profusely in the darkest ocean zone at about 2–4 °C, gives rise to, supports, and sustains biotic communities (Giongo et al., 2015).

Giongo et al. (2015) have extensively covered the chemosynthetic communities in Northern Atlantic and Pacific oceans. Limited investigations have been conducted on the African continent, as well as, the Latin American continent, specifically, Peru and Chile, and, also in New Zealand (Giongo et al. 2015). However, there was but scanty information concerning chemosynthetic activity in Western South Atlantic Ocean region until the investigations conducted by Giongo et al. (2015). The following are some of the striking features of the chemosynthetic communities as documented by Giongo et al. (2015).

Mat-forming vacuolated sulfur-oxidizing bacteria *Beggiatoa* (a genus of Gammaproteobacteria), *Azotobacter* (a genus of Gram-negative, spiral-shaped bacteria), *Thiothrix* (a genus of filamentous sulfur-oxidizing bacteria), and *Thiomargarita* (a genus of family Thiotrichaceae, which includes vacuolated sulfur bacteria species).

The common animal species existing amid chemosynthetic communities include polychaete annelids which belonging to the "Vestimentifera" clade which belong to the family Siboglinidae. Interactive relationships among various members of a chemosynthetic community involve direct predation as well as endosymbiosis.

It is important to note in this context, that, the ecological roles which the free-living chemosynthetic organisms and symbionts play, are of substantial biological interest, which needs to be thoroughly investigated, in the larger context of human welfare.

Hydrothermal Vent Ecosystems and Their Biodiversity

It is important to note, in this context, that chemosynthetic communities profusely exist on the floor of oceans throughout the globe. Prior to the discovery of the unique living systems thriving around the hydrothermal vents, on the ocean floor, in particular in the Pacific ocean floor and Galapagos Rift floor, around 1977, it was assumed that the exclusive source of energy for ocean life originated from the photosynthesis reaction above the land surface. Ecologists were completely taken by utter surprise at this unique discovery. In fact, this was nothing short of a "mystery of life". Hydrothermal vents could be very active, inactive, and sometimes, in between. Where there is underwater volcanic activity, on the ocean floor, high activity is seen in the hydrothermal vents. Fluids boiling at > 350 °C ooze out of the Earth's crust via cracks in the seabed, ejecting into the water column, through smoking chimneys. There is immense biodiversity, nourished by the chemosynthesis, in and around these active smokers.

It is in the warm water that chemosynthesis-based life exists, in utter darkness at the ocean floor, where sulfur compound is discharged from the hydrothermal vents. The sulfur compounds and the elemental sulfur form the source of energy for the bacteria to carry out their chemosynthetic activity. Elemental sulfur, hydrogen sulfide (H_2S) or thiosulfate present in the discharges of the hydrothermal vents undergo oxidation, and the energy so released is efficiently utilized by the sulfur-oxidizing bacteria in the process of carbohydrate synthesis, fixing carbon from CO_2 dissolved in water (Singh, 2020).

The following equations elaborate the process:

$$2H_2S + O_2 \rightarrow 2S^o + 2H_2O + \text{Energy}$$

$$6CO_2 + 6H_2O + 3H_2S + \text{Energy} \rightarrow C_6H_{12}O_6 + 3H_2SO_4$$

If one compares photosynthesis with chemosynthesis, it can be seen that in the case of photosynthesis, there is just one equation that represents the entire chemical reaction, while, in the case of chemosynthesis, there are several chemical reactions, based on the inorganic substance, as a source of energy.

The hydrothermal vent's biotic community, comprises of an abundance of sulfur-oxidizing bacteria, which are free living bacteria, and those dwelling symbiotically in the tissues of many of the invertebrates, living on the ocean floor. Of these, *Riftia pachyptila*, the "Giant Tubeworm" is one of the most thoroughly investigated invertebrates within the chemosynthetic hydrothermal community. A Giant Tubeworm can be as long as 150 cm in length and 4 cm in diameter. It has no mouth, no eyes, no legs, no alimentary canal, simply a very baffling creature that lives on ocean floor, entirely dependent on the chemosynthetic process for it's energy requirements. The "Giant Tubeworm" belongs to the family of Polychaete annelid Indian earthworms. It has a greenish brown spongy tissue, known as trophosome containing specialized cells which remain packed with chemosynthetic microbes—the sulfur oxidizing bacteria—which might number about 10 billion per gram of the trophosome and contributing 60% of the total biomass of the host. The plume, which is the upper part of this invertebrate, is reddish in color, owing to the presence of hemoglobin. With the abundant supply of hydrogen sulfide (H_2S) in the hydrothermal vents, and oxygen dissolved in water, the plume generates energy to utilize the carbon from the carbon dioxide (CO_2), in the marine environment, to synthesize biomolecules. It was the American microbiologist, Colleen Cavanaugh and her colleagues (Cavanaugh et al., 1981), who first discovered the chemosynthesis occurring within the trophosome.

The tube worm has a symbiotic relationship with chemosynthetic bacteria, and, it is this symbiosis, by which, the former nourishes itself. The outer covering of the giant tubeworm, made up of chitin, helps the animal survive extreme pressure and boiling-to-freezing temperatures on the ocean floor.

The tubeworm is, indeed, a keystone species, which plays a critical role in maintaining the structure of the chemosynthesis-fueled community. This community is protected from predators by the tube worm. However, several deep-sea crustaceans feed on the plume of the giant worm. Those predators which thrive on the tubeworm, hence, belong to the chemosynthetic community.

The chemosynthetic community dependent on the sulfur-oxidizing bacteria also thrive around the hydrothermal vents in deep freshwater, lakes, caves, as well as in hot springs. A hydrothermal vent community involving producers (free-living chemosynthesizers and those associated with the tube worm), and various levels of carnivores can be represented by an ecological pyramid, similar to an energy pyramid, discussed in the earlier chapters in this book.

From the time the deep sea hydrothermal vents were discovered in the Galapagos Islands in 1977, hundreds of other hydrothermal vents covering over five hundred animal species, surviving around hydrothermal vents, have been discovered. The major species include the following:

1. Eubacteria (Strain NS-E)
2. Archaebacteria (*Pyrococcus* strain GB-D)
3. *Thermococcus fumicolans*
4. Protists (Apicomplexa, Perkinsozoa, Syndiniales and Kinetoplastida)
5. Plant (Dandelions)
6. Animals (Tubeworms, Clams, Mussels, Crabs, Shrimp, Zoracid fish, Octopus etc.)

The submarine hydrothermal vents are biologically more diverse and productive. The thriving biotic communities represent a very high degree of species endemism.

An Ecological Pyramid representation of deep ocean floor chemosynthesis dependent hydrothermal community.

Top Consumers (Blind Crabs, Vent Octopus)

Secondary Consumers (Galtheid Crabs, Zoarcid fish)

Primary Consumers (Vent Zooplanktons, Vent shrimp, Vent Mussel, Giant Tubeworm)

Chemosynthetic Producers

(Vent Bacteria, Symbiotic Bacteria)

What Are the Threats to the Chemosynthesis-Based Biodiversity?

It is but natural that anthropogenic activities, cause dangers to the chemosythesis-based biodiversity, like other aspects of anthropogenic effects, on, such as soil resources, such as the huge soil degradation brought about by the chemically-centric green revolution (Nair, 2023a). The deep sea ocean floor hydrothermal vents are loaded with different mineral deposits. Some of these mineral deposits are of exceptionally high commercial value. And, these mineral deposits are exploited for profits by some of the private enterprise. The deep sea mining industry is just in its infancy, but, its potential to adversely and effectively damage the rich biodiversity is only a matter of time. Many private companies, with expertise in deep-sea mining have already won contracts to exploit valuable mineral wealth deposited within the living communities, on the ocean floor. An active hydrothermal vent spreads to just about 50 km^2 and many activities on or near an active hydrothermal vent would invariably lead to the destruction of the living organisms within its ambit. Losses of biodiversity would be huge and irreversible, due to the total alteration/destruction of the living communities.

In addition to the above described threat posed by mineral extraction, another threat arises from the pharmaceutical industries/interests which are looking for newer substances having unique pharmaceutical properties. Some private enterprises are already well advanced in the exploitation of the rare species for manufacturing newer products of pharmaceutical importance and extraordinary socioeconomic value. Once, the large-scale exploitation of the mineral resources and living communities on the ocean floor takes place, the ecosystems blossoming like oases on the ocean floor, would be in a very perilous condition. It will be a similar case like the "LPG Agriculture", where monocropping, monopolized by three crops, namely, rice, wheat and maize, and its private ownership by MNCs like Monsanto and Bayer, will lead to the ultimate destruction of all agrobiodiversity. Until now, it was believable that deep-sea creatures would be safe from the ongoing Sixth Mass Extinction against the backdrop of Anthropocene. Not any more. It is now observed that even the seemingly unseen, invisible species of life, which have opened up a huge door for evolution through their chemosynthetic processes, are not free of the human-lead greed culminating in total extinction of all life.

Words International Obligations

One observes that hydrothermal vents are found, quite often, beyond national boundaries, because of the connection of seas between countries and continents. The United Nations Law of the Sea Treaty has established an organization known as International Seabed Authority (ISA) to look into all questions arising out of territorial dispute, as well. The ISA has the complete authority to regulate all laws concerning the mining of the hydrothermal vent zones. Global attention is centered on the exquisite value, and, rarity, as well as, the vulnerability of these vent zones. Parties of the Convention on Biodiversity (COB) held in Rio de Janeiro in 1992 put in place the recognition to hydrothermal vents as "ecologically and biologically significant areas" in need of enhanced conservation and management measures. The UN General Assembly has given a call to "manage all the risks to the marine biodiversity of hydrothermal vents".

It is very important to note that the chemosynthesis-dependent biodiversity is the last remaining component of the relatively undisturbed biodiversity, which is not easily accessible to human exploitation. This would mean that this important component deserves the maximal protection from human intervention to meet man's greed. In fact, the hydrothermal vent biodiversity is Nature's "virgin biodiversity". This also implies that the history of evolution of life is hidden inside this "virgin biodiversity". It is yet to unfold fully. Once this "virgin biodiversity" is disturbed or damaged, due to human greed, the biodiversity blossoming at its ecological climax, amidst deep-sea active hydrothermal vents will completely perish. And with this, all the mysteries of Evolution of life, will vanish for ever, from the face of planet Earth. We, as conscientious scientists, have a moral obligation to not let this happen, because, the generations to come will never forgive us for this blatant thoughtlessness.

As conscientious scientists, it is very important for us to fully realize that the immediate protection and conservation of hydrothermal vents, the "virgin biodiversity" on planet Earth can simply be achieved simply through laws, alone, even if internationally binding. Human greed cloaked as "private initiative", will always find means and methods to circumvent this safeguard. That has been the history of the 21st century man! The global climate change scenario is the best living example of this "circumvention". Despite innumerable Conference of Parties (COP) on climate change, going on since decades, it has not been able to cap the rise of global warming, which has led to the warmest years on the European continent, precisely more than 2 °C, on the European continent, leading to the warmest years (year passed) ever in human memory, since writing this book. Like climate change, it will always be the economically weakest and poorest nations on the face of the Earth who will pay a huge price, in terms of lives lost. This "virgin biodiversity" and the ecosystems infusing life on planet Earth, should be well guarded, and, kept away from human mindless exploitation for the so-called "economic purposes", or, what is these days, fashionably, called "economic progress". It is not only necessary, but, also imperative in our own life to halt this kind of thoughtlessness. Not one Greta Thunberg (the teen-aged climate activist, from Sweden, who set sail on a boat, ignoring an air flight, from Stockholm to New York, to attend the global summit on global warming for the benefit of all Heads of State, organized by Mr. Antonio Guterres, United Nations Secretary General in New York on September 23, 2019, who hit the headlines a couple of years ago), but, only thousands of such Thunbergs, who can stall this plunder of the ocean bed. It is the supreme obligation on the part of humanity, so that, this generation does not "gift" an totally ruined environment to the future generation, as has happened with the highly chemical-centric "green revolution" (Nair, 2019a, 2023a).

References

Arvidson, R. S., Morse, J. W., & Joye, S. B. (2004). The sulfur biogeochemistry of chemosynthetic cold seep communities, Gulf of Mexico, USA. *Marine Chemistry, 87*, 97–119.

Cavanaugh, C. M., Gardiner, S. L., Jones, M. L., Jonnasch, H. W., & Wartrbury, J. B. (1981). Prokaryotic cells in the hydrothermal vent tube worm Riftia pachyptila: Possible chemoautotrophic symbionts. *Science, 213*(4505): 340–342. https://doi.org/10.1126/science.213.4505.340.

Giongo, A., Haag, T., Simao, T. L., Medina-Silva, R., Ultz, L. R., Bogho, M. R., Bonatto, S. L., Zamberlan, P. M., Auguastin, A. H., Lourea, R. V., Rodrigues, L. F., Shrissa, G. F., Kowsmann, R. O., Freire, A. F. M., Miller, D. J., Viana, L. F., Ketzer, J. M. M., & Eizirik, E. (2015). *Discovery of a chemosynthesis-based community in the Western South Atlantic Sea.* Deep Sea Research Part I. https://doi.org/10.1016/i.dsr.2015.10.010

Kaim, A. (2010). Chemosynthesis-based communities through time. *Geophysical Research Abstracvts*, 12 (EGU2010-12679-1).

Lindeman, R. L. (1942). The trophic-dynamic aspect of ecology. *Ecology, 23*(4), 399–418. https://doi.org/10.2307/1930126

Milucka, J., Ferdelman, T. G., Polerecky, L., Franzke, D., Wegener, G., Schmid, M., Lieberwirth, I., Wagner, M., Wideel, F., & Kuypers, M. M. M. (2012). Zerovalent sulphur is a key intermediate in marine methane xifdation. *Nature, 491*, 541–546.

Nair, K. P. P. (2019a). *Inelligent soil management for sustainabale agriculture: The nutrient buffer power concept*. Springer Nature, Switzerland.

Nair, K. P. P. (2023a). *The living soil: A lifetime journey in understanding it for human sustenance.*. SpringerBriefs in Environmental Science: Springer Nature.

Singh, V. (2020). *Environmental plant physiology: Botanical strategies for a climate smart planet.* CRC Press (Taylor and Francis). ISBN: 9780367030421.

Chapter 5
Which Are the Threats to Biodiversity? It's Conservation and Sustainability

It is a very vexing scientific question, and, that concerns the severe threats biodiversity is now facing globally. When a very valuable species is threatened, it has a chain reaction. Many species have become extinct and many more are soon to follow this fate. The former chapter (Chap. 4) has elaborately discussed the threats which the hydrothermal vents, on the ocean bed, are now facing globally. The effect of this, is that, a very valuable process of synthesizing energy, the chemosynthesis, unlike the photosynthesis, is being now threatened by human greed.

Primarily, it is the anthropogenic (human) factors that pose the greatest threats to biodiversity. Human activities, simply put, are malovalent to the flourishing of biodiversity. The primary cause for the destruction of global biodiversity is the ever growing human population. But, at the root of this threat is human greed. Taking the example of the chemically-centric green revolution *vis-a-vis* the "LPG Agriculture", one can imagine to what extent human greed can go—enriching the few rich at the expense of the vast many poor and unprivileged. The MNCs in seed business, like Monsanto and Bayer, have controlled the global seed business and confined them to just three crops—wheat, rice and maize. Many others are just thrown out of the window.

Natural evolution is never a threat to global biodiversity. A species at a specific point in its evolutionary cycle, is destined to vanish, at some time or the other, but, the void created due to this loss is filled by the evolution of new forms. Not only is this void filled up, the Nature, gradually, but steadily, goes on enriching itself with more life, in tune with the pace of evolution. It is a blessing that the biosphere has become vibrant in its evolutionary journey. However, this evolutionary journey is cut short by the anthropogenic factors, to put it blankly, the "human greed".

K. P. Nair, *Biodiversity in Agriculture*,
https://doi.org/10.1007/978-3-031-44252-0_5

How Are Habitats Destroyed?

The primary cause or the destruction of biodiversity, originates, in the first place, the destruction of habitats. The habitats are the abode for all living species and their diverse genotypes. Currently, in the ecosystems, there are, but limited number of habitats. There is hardly a "virgin forest" on planet Earth. The forest wealth has been plundered ad infinitum by the vested interests thieves. Forest ecosystems have been ruthlessly converted into cultivated land, grazing lands, grasslands, and mining sites. What was once a habitat of wild animals is now human "settlements"—villages towns, cities, commercial and industrial complexes, tourist destinations, playgrounds, honeymoon huts and such others. Those in power who establish such avenues have no regard to what the long-term implications of their decision would be on biodiversity. The only interest is to generate more money, no matter what happens to the biodiversity. Some of the "conversions" have led to terrible ecological disasters. When habitats which provide a refuge for a variety of flora and fauna are destroyed, indiscriminately, for human use—to better use the term to meet "human greed—wild animals and endemic flora suffer in different ways, sometimes in irretrievable ways. They end up as "ecological disasters". But, happily, and very encouragingly, there are some exceptional instances in the history of biodiversity conservation, as well. For instance, the "Silent Valley Agitation", in the State of Kerala, India, is a shining example. The Silent Valley in Kerala State is home to one of the richest biodiversity habitats in the world. Several decades ago, there was a move by the State Government of Kerala, to clear the dense jungle and erect hydroelectric stations. It was the people of Kerala, at the grassroots level, who took up the cudgels against the covert attempts by those in power in the State Government. After a prolonged battle/agitation/sit-ins etc., the people won, at the end, notably because of the perseverance of their agitation against the governmental moves. Laudably, after some years, at the end, it was the direct intervention of the then Prime Minister of India, late Indira Gandhi, who intervened to stop the project.

Now, the Silent Valley remains as one of world's last tracts of undisturbed tropical evergreen forests across the world.

Another notable movement is the "Chipko Movement" in northern India, led by the late and illustrious socialist leader Jayaprakash Narayan, where the poor villagers hugged trees from being felled. Trees stood for the "5 Fs"—Food, Fodder, Fiber, Fuel and Fertilizer.

Both the Silent valent Valley and Chipko Movements were a great success stories in the world, in terms of biodiversity conservation.

When habitats which provide refuge to a huge collection of flora and fauna are destroyed, also when they are transformed for other purposes, for instance, human use, endemic flora and wild animals suffer in many ways. The displaced animals do not get sufficient food, food chains break down, disruption of breeding ground for wild animals takes place, scared animals will have no other option, but, to survive under great stress, the breeding capacity of animals decline, routes of migratory animals disrupted, native plants do not grow and prosper following their natural vigor.

Symbiotic relationships, among different plants, animals and even microorganisms, are severely broken down/disrupted following habitat destruction. The species population goes on decreasing and the net result is that the modified environment leads to the extinction of many species of plants and animals.

How Does Habitat Fragmentation Take Place?

When a huge intact habitat is split into two or more fragments, habitat fragmentation is considered to take place. Fragmentation of habitats is an inevitable reality of our current times. This is due to the several "developmental" activities which involve large habitats, for instance, building of roads, railway tracts and so on. The word developmental has been put under inverted commas, to signify, that quite a few times, such activities are not necessary. One classic example is of the Kerala Government proposing to build a high speed train between the State's capital city, Trivandrum, and the end of the geographical limit of the state, Kasaragod, and, is called as the "Silverline Project". On a careful scrutiny of this project, by this author, it was found that the envisaged/proposed project would lead to several ecological disruptions. The project would have involved dislocations of several thousands of residents of the State, throwing life and property into disarray. There were several demonstrations against the project by the people of Kerala. At the time of writing this book, the word is that the project is now shelved.

Habitat fragmentation disrupts life and biodiversity in many ways, the following are some:

1. Limiting dispersal and colonization potential of species
2. Reducing the foraging ability of wild animals
3. Disturbing ecological niches and/or separating species from their niches.

The third point is that of great concern. The Silverline project would have cut through the vast wetlands of Kerala State, namely, the *Kuttanad* region, which is the rice bowl of the State.

Exotic Species—Their Introduction

In a common environment, all the species and their genotyes co-evolve, over millennia, which lead to the establishment of interrelationships and interdependence, synergism and symbiosis. A community generally does not allow an alien or exotic species to enter its boundary (ecosystem), unless some adverse factors prevail in the community. As a rule, most of the exotic species also do not often prefer to establish in a new community. However, if a specific species establishes in a new environment/community, it would spell disruption among the existing species. The introduction of an exotic biota can be both accidental and/or intentional. In both of the situations, the ecological consequences can be very deleterious. In a new community, an exotic/introduced species can exert pressure on a new community in several ways and may lead to serious ecological and environmental consequences. The following are the examples of some of these after effects:

1. The structure of the ecosystem can be altered, also the composition of the species
2. Inductment of habitat can lead to alterations to the extent that native species would no longer be able to persist
3. The release of allelochemicals which are detrimental to many native species and the existing soil microflora
4. Preying upon native wild species leading to total extinction
5. Leading to the incidence of epidemics, for instance, in the case of some microorganisms, leading to elimination of native species
6. Acting as causal agents of some of the communicable diseases giving a jolt to public health
7. The functional attributes are dramatically affected, in an ecosystem, such as biodiversity conservation, photosynthetic efficiency, nitrogen fixation rates, finally affecting the primary and secondary production rates.

The ecological effects of exotic species on island ecosystems—which are home to most of the global rare and threatened species are especially very grave.

Once when an exotic/introduced species colonize in a specific community of species, the exotic species will, without doubt, replace the endemic species.

Indeed, there are some very practical examples, where some of these exotic plant species have turned to be a global menace, and the following are some:

Parthenium hysterophorus: Native of tropical America, called Santa Maria Feverfew introduced into India, post green revolution, a weed in wheat fields. It is also known, commonly, as famine weed.

Eichornia crassipes: It is the native of South/Latin America. Popularly known as water hyacinth, an aquatic plant, which proliferates hugely in freshwater ponds and rivers, clogs rivers. One of the fastest growing aquatic weeds, which threatens the survival of many other aquatic plant species.

Eupatorium perfoliatum: Known as common boneset. Introduced into India through US food import (PL 450 Food Aid of the sixties).

Lantana camera: Originated in West Indies, has become a pernicious weed in India. Flowers throughout the year and has very high fecundity. The plant has ornamental value, introduced by the British into India, but, has become a pernicious weed in all places in India.

Among the most cited examples of animal alien species is the Nile perch introduced in Lake Victoria of South Africa. *Lates niloticus*, commonly known as Nile Perch, a fresh water fish and of very great economic value to East Africa, a native of the Nile, Congo, Niger, Senegal and many other river basins in the Afrotropical realm was introduced to Lake Victoria in the fifties. This alien predatory fish threatened the entire freshwater lake ecosystems eliminating many species of endemic Cichlid fish, a freshwater fish, mainly found in tropical America, mainland Africa, Madagascar and South Asia. When this fish was eliminated, the Nile perch began preying upon small shrimp and minnows.

The Hazards of Hunting and Overexploitation

Humankind has the hunting habit since time immemorial. It is the predatory nature of humankind/man. In fact man was known as a hunter and gatherer since millennia. The habit of hunting is not merely to fulfill essential needs or to take safety measures, but, also to satiate a will by posing as a triumphant. Killing of animals, by the "art" of hunting, for economic gains, taste, pleasure, recreation etc., are the other objective of hunting. The two most striking examples of the consequences of hunting are the following:

1. Disappearance of the Cheetah (*Acinomyx jubatus*), India's fastest mammal
2. Disappearance of the Dodo, (*Oidus impetus*) a unique bird of Mauritius.

The overexploitation of natural resources wantonly to satiate economic greed of man is leading to the disappearance of several plants and animals, of great importance to global biodiversity, from the terrestrial regions and oceans.

The Hazards of Poaching and Smuggling

Although possession and trade in the body parts of animals is legally prohibitory, man resorts to such cruel acts to satiate his pomp. The classic case is of illegal possession of tusks of elephants, and, other ivory products. Several animals in their wilderness are being killed for their products be it tusks as in the case of elephants, or skins, as in the case of tigers. Notorious poachers and smugglers manage illegal trade with the help of the mafia. Such malpractices pose grave danger not just to the animal species, but, the plant species, as well.

The Hazards in Agriculture

Agriculture is the largest land-based human activity, involving crop husbandry, forestry, horticulture, olericulture, floriculture, animal husbandry, dairying, poultry, etc., that has devoured a huge chunk of planet Earth's biodiversity. Agriculture is not just the act of producing food for human consumption, and, most of the vital needs of humanity, but, also an activity compromizing the biodiversity of Nature. The percentage of biodiversity used in managing agriculture is referred to as agro-biodiversity. Agrobiodiversity encompasses significant geographical space on planet Earth's surface, but, only the tip of the global biodiversity. Only a few plant species are grown for food production. Amongst them, only three species, wheat, rice and corn/maize, occupy the largest proportion land on planet Earth, about three-fourths of human energy needs. We hardly encounter more than two dozen plants of cereals, vegetables, fruits, beans etc. providing us food. For clothing needs, we exclusively depend on cotton. There are some other plant species, which can also be cultivated to meet human need. But, the area occupied by these plants/crops is very minimal. Animal husbandry involves just a few species of animals, such as bovine, ovine, canine, and equine etc.

The most tragic fate that agrobiodiversity faces on planet Earth is the, almost, total, extinction of genetic biodiversity. When we go back, to millennia, especially, when man started tending plants/crops in a limited space, than going hunting and gathering, which later came to be known as the "agricultural revolution", we find that numerous varieties of crop and/or plant species were grown. In the so-called modern agriculture, or what was euphemistically known as the "green revolution"—a highly soil extractive farming, which comes under the umbrella of "LPG Agriculture", the genetic base of food crops carries no value. The only target in cultivating a crop is "productivity". In order to chase the maximal productivity, only a few varieties, known as "High Yielding Varieties" (HYVs) or "High Responsive Varieties" (HRVs "High Responsive Varieties"—highly responsive to external inputs, such as, unbridled use of synthetic fertilizers, copious irrigation water and an array of pesticides and herbicides—all external inputs) were cultivated. In fact, behind this move, was the seed and fertilizer business. In animal husbandry practices, only a few breeds of cattle, buffaloes, goats, sheep, horses, camels, and poultry birds were left back on planet Earth. The consequent extremely narrow genetic base has turned out global agriculture into an extremely vulnerable sector. Narrow species and genetic bases invite more on slaughts of pests and epidemics. This has been the fate of the so-called green revolution of the HYVs and HRVs. Look at the total destruction of rice crop in Asia due to the attack of Brown Plant Hopper (*Nilaparvta lugens* (Stahl).), look at the widespread destruction of the wheat crop due to wheat rust, leaf and yellow rust, (*Puccinia triticina, P. stritiformis* f.s.p.tritici) in India. This state of agrobiodiversity management imposes the use of very expensive, but, deadly pesticides and herbicides, which in the long run has created a trail of soil, water, air and food pollution.

The Hazard of Environmental Pollution

Pollution is disruptive in character. It's role is life-annihilating. Look at the consequences of unbridled use of pesticides and herbicides in Punjab State, India, especially in Gurdaspur, in Punjab State, the "cradle" of Indian green revolution, which has now become, thanks to the green revolution, the "capital of cancer". Pollution plays a life-annihilating role on soil, water and other terrestrial ecosystems. Look at the ground water pollution, post green revolution—the excessive concentration of chemical residues, which has made ground water no more potable in Punjab State.

Pollutants in terrestrial ecosystems poison food chains, and, also terminate sensitive species—both plant species and animal species. Oil spill changes the quality of sea habitats. Heavy metal pollution in water systems accumulates in the bodies of organisms and kills many sensitive ones. Many predatory birds, feeding on fish from polluted water bodies die due to the biomagnification of the toxic pollutants.

The Cycle of Man-Wildlife Conflict

One of the most important factors to take note of in biodiversity discussions is the perpetual cycle of man-wild life conflict. It is, but natural, that wild animals tend to migrate from one region to another, for several reasons, such as, availability of food, shelter and to be free of the danger of the man-wildlife conflict. The animals are, more often than not, dynamic in that they often break the boundaries of their habitats, even if there is no paucity of food. However, in almost all of the cases, wild animals cross their boundaries when they sense and/o face a shortage of food of their natural choice in their own habitat. As human intervention in wildlife habitats is increasingly intensified, a state of ecological disaster is unfolding to the extent that wildlife is forced to migrate, and, roam around human settlements, and, attack human beings and their livestock. The newspapers carry frequently news of animals attacking and killing human beings and domestic animals in the human settlements near protected areas. In response, local people also kill the attacking animals. Thus, a man-wildlife conflict goes on perpetually. Even the herbivorous animals which do not attack human beings, are killed because they destroy planted crops. When an animal gets injured by a human being, it retaliates. Even a weak or infirm animal also develops a tendency to attack human beings. Most of the time, female animals attack human beings or other animals, who pose a threat to their infants. If a lion, tiger, leopard or any other carnivore animal tastes human flesh, it becomes a man-eater. Though it might be difficult, it is not entirely ruled out, that, a human being can kill an animal if it turns out to be a man-eater. But, in the process of tracing a man-eater, many innocent carnivores become a prey to human fury of revenge.

Many a time, wild animals' migratory routes are destroyed when construction of roads, bridges etc., takes place. Consequently, the displaced animals tend to intrude into human settlements. And, often, make the humans, easy prey for themselves. However, in this cycle of man-wild life conflict, it is the animals who suffer the most.

Currently there is hardly any provision for a fair compensation to the famer, when his crops are damaged by intruding wild animals. The problem is, the laws are very lax on this aspect. Ultimately, it is the farmer who bears the brunt of financial loss. Also, there are no fool-proof methods by which efficient crop protection can be ensured from wild animals. This author is aware of some cases where electrocuting devices are erected to ensure safety of the crops from wild animal attacks. However, all these measures are either temporary or not fully fool-proof. This leads to the local farmers getting rid of wild animals by indiscriminate killing.

How to Avert Man-Wildlife Conflicts?

1. Ensure that there is food, fodder, and water within protected areas and also in natural forests, aplenty, the most likely effect will be that animals will confine themselves to their habitats and would not cross over to other places of human habitation, in search of food.
2. Alternative cropping patterns may be followed by farmers, involving crops which are unpalatable to wild animals, especially for wild herbivores. Hence, animals would not venture to consume the crops which are not to their liking.
3. Ensure that migratory routes of wild animals are neither disrupted nor obstructed. Appropriate corridors should be provided to facilitate the seasonal migration of wild animals.
4. It is advisable to raise dense bio-fencing, involving specific trees and bushes which can be raised on the boundary of villages, near national parks and wildlife sanctuaries. Similarly, ensure erection of these tree and bushes on the boundaries of protected areas as well.
5. When herbivore animals destroy/consume crops of farmers, there should be adequate provision to compensate the economic loss incurred by these farmers.

Species Extinction

Among the most serious consequences of the biodiversity conservation on planet Earth is the question of the extinction of the species. It is important to remember that once a specific species becomes extinct, the most valuable genetic information contained in its cells, the DNA, is also lost forever. It is also important to note that once a specific species becomes extinct it is also at the cost of many others to follow. Also, the possibility of its further evolution will be lost forever. Once a specific species becomes extinct, many others dependent on it, will also near extinction.

How shall one decide/conclude that a specific species is extinct? This happens, when the last known member of that species, anywhere in the world, is dead and/or destroyed. This is tragically the case with modern agriculture. It is against this backdrop that Nair (2019a, 2019b, 2019c) has made out a strong appeal to search, catalogue and utilize the Crop Wild Relatives of many of the current species of plants grown for food, fiber, fruit or vegetable. There are many members of many species, which are no more alive in the wilderness, but, sadly, only in captivity, that is under controlled conditions, imposed by man. Such species are said to be extinct in the wild. Under both conditions, such species are considered to be globally extinct.

There is also another scenario, where it represents a species, but, left in such a small measure, that its impact on the community of species that it exists in, is but, very negligible. These species can be said to be ecologically extinct. A case is, where, the number of tigers left over in the wild is so very negligible that their effect on the prey, that is herbivores, is next to nothing.

Inasmuch as ecosystem functioning is concerned, a climax species plays a significant role, in water conservation, microclimate development, and ecosystem productivity etc. However, on account of human intervention, the species number is reduced to such an extent that its phenomenal contribution to ecosystem functioning will be next to nothing. Such a situation is termed as "functional extinction."

What Are the Causes/Reasons for Extinction?

In nature, the extinction of species in not a new phenomenon. If one considers evolutionary history of life in the Nature, species have been done with and new species have evolved. About 570 million years ago, multicellular organisms evolved on planet Earth. As per some estimates, as many as 30 billion species have thrived on planet Earth during the entire history of biological evolution on the Living Planet. As much as 99.99% of all species, which evolved over time, at some point in the very long geological scale have become extinct. Poignantly, it can be said that extinction is the ultimate fate of any species, mankind included. The natural extinction, which is the extinction of a specific species in its millions of years long journey, has to inevitably meet, which has never been a concern of human worries. In recent times, human activities (anthropogenic in character) resulting in environmental disruption and destruction, have led to accelerated extinction, at an alarming rate. This, indeed, is a worrisome state of affairs, stemming from human greed. There are, on the whole, three processes by which the species met the fate of their extinction, and they are the following:

1. Natural Extinction
2. Mass Extinction
3. Anthropogenic extinction (rapidly contributed by greedy human activities).

What is Natural Extinction?

It is a fact of Nature that every individual of every species on planet Earth will be extinct at some time or another. It would die as per the natural death program, known as apoptosis, they are born with. However, species, on the other hand, live for a very long time, unlike of course, man's greed destroys it. But, it does not live forever. An individual species has also to meet with its end in response to some laws of Nature, not fully known, to date. The death of a species with all its individual members and the genotypes, from everywhere on planet Earth in a natural manner is termed natural extinction. It is generally thought that some species disappear and others "more adapted to changed environmental conditions" replace them. Singh (2019) in his theory of "Fertilizing The Universe", ascribes this phenomenon to the role—light-converted-to-biochemical energy plays in living energy systems. There is another term that describes the natural extinction that has occurred in the geological past, which is, referred to as background extinction.

What is Mass Extinction?

Extinction of large number of species has taken place at various stages of geological history, and, this arises out of global catastrophes. When the rate of extinction or disappearance of the species is faster, more rapid, compared to natural extinction, or background extinction, over a geological scale, mass extinction is supposed to have taken place. Mass extinction has taken place over millions of years. It is estimated that more than 65% of all species on planet Earth met with their fate of extinction due to global catastrophic events over a few million years, which is a period much shorter compared to the history of planet Earth, spanning over some 4.6 billion years. About 225 million years ago, the most severe mass extinction leading to the disappearance of over 95% of the existing marine species took place. About 65 million years ago, the most discussed, and the last mass extinction, which wiped out completely the giant dinosaurs, ruling over planet Earth for as many as 140 million years, took place.

Both environmental and biological factors caused the catastrophical global mass extinction. The catastrophical factors/events, are global cooling (as opposed to the current oppressive global warming) a decrease in sea level, predation and competition. Different epochs of global cooling brought the planet Earth's ambient temperature down to a level, where most of the existing species could not maintain their physiology, and, hence, they vanished. The global cooling must have triggered the lowering of the sea level. Most of the water that evaporated froze and turned to ice and snow, especially on the poles and the mountains leading to a significant fall in sea levels. Unusual rates of predation and competition, among species, might have resulted due to ecological changes induced by the adverse thermodynamic conditions of the biosphere. A recently proposed theory concerning mass extinction, especially that relates to the dinosaurs, proposes that a giant comet or an asteroid hit planet Earth, which led to the total wipe out of masses of species directly, owing to the huge clouds of dust which prevented the rays of the sun reaching the planet earth, for a very long time.

What is Anthropogenic Extinction?

Man's activities, most of the time, due to his greed, is referred to as anthropogenic, leading to anthropogenic extinction. Disrupting ecological systems/processes, over—exploiting natural resources, and annihilating living species, unabashedly and indiscriminately, which are of no direct consumptive value and/or are thought to be undesirable or harmful. The current rate of extinction is 1000–10,000 times more than the background or natural extinction. The following list, finalized by the International Union for Conservation of Nature (IUCN) is, indeed appalling:

1. About 4500 species of mammals, birds and amphibians are at the grave risk of mass extinction
2. The above number includes 1200 mammalian species, 1300 avian species, and 2000 amphibian species

3. Of all the described species, above, the proportion of the species deemed to go extinct is 26% mammalian, 13% avian, and 41% amphibian
4. With the current rate of the loss of species, we may witness the loss of nearly half of all living species by 2100.

How Susceptible Are Species for Extinction?

All living species are not equally vulnerable to extinction. Some of the specific characteristics of each species make it more susceptible than the others. The following are some of the examples:

1. Feeding at a high trophic level: It is seen that with regard to the energy trapped by the producers in the ecosystem to sustain life, the top carnivores are left with only the minimal amount. On the other hand, herbivores, the closest ones to the producers, on the other hand, have plentiful of food/energy to sustain their life. This would imply the following:

The further a specific species is from either photosynthesis or chemosynthesis, the more susceptible to extinction, it is. For example, the Bengal (Indian) tiger and the Bald eagle are the main species facing extinction because of top carnivorous habits.

2. Low fecundity coupled with small population: Those species which have these characteristics are in a constant state of pressure from other species which has a higher fecundity and larger population. The Blue Whales and the Giant Panda are strange examples, which face such vulnerability.
3. When distribution is localized and which has only a narrow range: When the geographical area in which a specific species thrives, if it is smaller and narrower, the chances of a species, being vulnerable to extinction, is quite high.
4. Fixed migratory routes: Species such as the Blue Whale and the Whooping Crane, which follow fixed migratory route, are quite likely to be at a much greater loss, since they encounter several obstructions, imposed by adverse environmental conditions and enhanced chances of becoming victims of other predatory species.
5. When the body size is large: Species such as, the elephants, the Bengal (Indian) tiger, lion, are a lot more likely to face extinction, because of their huge requirements of ecological space and food on account of their large body size.

The Red List or Red Data List of threatened species, globally, by the International Union for Conservation of Nature (IUCN), is the most comprehensive inventory of biological species. Founded in 1964, the Red Data List comprises the taxa faced with the severe risk of extinction, which applies to all species and all regions of the world. Following are the main objectives of the IUCN Red Data List:

1. Identification and documentation of all threatened/endangered species across the globe
2. Providing the Global Index of the decline in biodiversity

Table 5.1 Details of the IUCN Red List categories

Details of the Red List category S. No.	A taxon is placed in the category when it meets with the following conditions
1. Extinct (EX)	When there is no reasonable doubt that the individual species is dead
2. Extinct in the wild (EW)	When it is known to survive only in cultivation, captivity, or as a naturalized population, well outside the past range; exhaustive surveys in known and/or expected habitats fail to record the presence of an individual
3. Critically endangered (CR)	When it is facing an extremely high risk of extinction in the wild in the immediate future
4. Endangered (EN)	When it is not critically endangered, but, is facing high risk of extinction in the wild in the immediate future
5. Vulnerable (VU)	When it is not critically endangered, but is facing a high risk of extinction in the wild in the medium-term future
6. Near threatened (NT)	When it is evaluated against the criteria, but, does not qualify for a threatened category in the near future
7. Least concern (LC)	When it has been evaluated against the criteria but is not found qualifying for CR, EN, EN, VU, or NT (indicative of widespread and abundant taxa)
8. Data deficient (DD)	When there is inadequate information to make a direct or indirect, assessment of risk of its extinction
9. Not evaluated (NE)	When it has already been assessed

Note Individual countries and/or organizations also prepare the Regional Red Lists which are helpful in assessing the risk of extinction of species within different political management systems
Source IUCN (2019)

3. Development of awareness about the importance of endangered/threatened species (biodiversity)
4. Defining conservation priorities at the local level and guiding the conservation action.

In Table 5.1, the nine Red List categories of the species defined by IUCN (2019), according to the degree/extent of the threat perception, are listed. Each taxon of the biosphere, except microorganisms, can be classified into the following categories listed in Table 5.1.

Biodiversity Conservation

It is a tragedy of our times that, if not all, but, almost all of the ecosystems which are on planet Earth, have ceased to be virgin. Anthropogenic designs have modified them beyond recognition. The blossoming of a climax vegetation is, indeed, a rare sight. As human intervention into the Nature has escalated, uncontrolled, all types of biodiversity—genetic, species and community, also, alpha, beta and gamma—have been under stress. Much of the biodiversity has been lost from the face of planet Earth, due to, what I may call "anthropogenic greed", further aggravated by the climate crisis, which, again, is caused by "anthropogenic greed" (Nair, 2019a, 2019b, 2023a).

The biodiversity is being lost at a rapid rate and is continuous. This is, indeed, a serious ecological threat to life on Planet Earth. This is dismal and this will not enable us maintain a sustainable future. It is incumbent that the current generation hands down a sustainable environment to the future ones. The planet Earth belongs to our children more than to us. Conservation of biodiversity, hence, is a great matter of ethics. Ethics is the human face of biodiversity.

I suggest the reader to imagine how the world would look like without any biodiversity. It certainly won' look infatuating, indeed, it would look so very barren and bleak. Beauty is an expression of life-supporting, life-sustaining and life-enhancing environment. In fact, beauty emanates from the richness of life, not from lifelessness. Embodying the richness of life, biodiversity, hence, spells out it aesthetic values. The coming generations inherit the right to live amidst a beautiful environment of blossoming beauty, not something like a dead desert! Conservation of biodiversity, hence, is a must with top priority for the sake of ethical and aesthetic values.

It is an irony of life, and, of Nature in totality that biodiversity and its unique importance has neither been totally and completely well understood, nor, does it seem possible to understand. Many of species have become extinct as they were supposed to be. Several undesirable lands, referred to as "weeds" vanished due to human intervention. But, how can one be totally certain whether these so-called "weeds" are exclusively valueless or they have some inherent value, even it be potential? These may be absorbing some optional values, which we could not capture, to date, and those optional values might prove very vital for human welfare, in the future. Hence, species must be saved/preserved and conserved for the sake of their optional values, which may open up new windows of our knowledge, about the plants.

There is no single species on planet Earth that can be discarded as "useless". Every species came into existence as a result of natural evolution. How can then a species be useless or valueless? Every taxon in existence is laden with valuable existence value. Existence value is a vital attribute of natural evolution. Biodiversity must be conserved to recognize and uphold the sacred existence value of all life forms, and all species and for appreciating natural evolution and marveling at its living attributes, adorning the planet Earth, with a wide diversity of life. The beauty of planet Earth lies in its biodiversity. In other words, biodiversity is the natural foundation of the beauty planet Earth embodies.

Biogeochemical cycles, water cycles, weather cycles, seasons and climates, all together contribute to an ecological balance very vital for the biosphere to effectively maintain its life-giving, life-supporting and life-enhancing functions (Singh, 2019). It is just that the "anthropogenic greed" that disrupts/disturbs this ecological balance.

The human race is morally and ethically bound with an inalienable obligation to protect, regenerate and enhance natural resources, with the richness of biodiversity. The conservation decisions human race takes to day would, needless to add, have a phenomenal impact on the future biodiversity, in fact, the very quality of life on planet Earth.

What Are the Strategies for Biodiversity Conservation?

When sound management principles are accepted and implemented, concerning all living species, so that all living species and their genotypes are held in safety for use by the current, as well as future generations, one can expect good conservation strategies to have been implemented. At the forefront of biodiversity conservation, the first and the foremost principle is, the halting of any further deterioration, physically or otherwise, of the existing and thriving species—of animals and plants. Prevention of deterioration, degradation, depletion and degeneration—the four "D"s—must be rigidly ensured. This would be, when the guardians of biodiversity conservation can prevail upon the coterie of "anthropogenic greed". However, there are two major challenges, enumerated below, in implementing thoroughly, conservation strategies:

1. Limited availability of space and human "pressure" (exemplified by the anthropogenic greed) on natural resources.
2. The two basic biodiversity strategies applied are

   A. In-situ conservation
   B. Ex-situ conservation.

Both of the above have their own advantages and disadvantages from an implementation perspective. Each is explained below.

In-situ Biodiversity Conservation Strategies

In-situ (in it's own place) conservation: It is the method of conserving all the living species, especially, the wild and endangered species, in their natural habitats and environments. In-situ conservation of biodiversity includes biosphere reserves, national parks, wildlife sanctuaries etc.

Ex-situ Biodiversity Conservation Strategies

Ex-situ (out of it's place) conservation: It is the method of conserving all the living species in an artificial habitat. Examples include aquariums, botanical gardens, cryopreservation (in the laboratory) and DNA banks etc.

In-situ Strategies of Biodiversity Conservation

All the living species which prosper on Planet Earth have evolved in the natural environments, which are appropriate for their evolution, and, where they can adapt along with other associated species in harmony with the operating environmental factors, conducive for their survival, functions and reproductive activities. Providing due protection to the entire ecosystem where communities of diverse species prosper is what the in-situ conservation method calls for. For the operationalization of this strategy, a specific group of ecosystems is protected and thus, a network of protected areas is earmarked for biodiversity conservation.

What Are the Protected Areas?

These include earmarked and defined geographical areas especially dedicated to the protection and maintenance of biodiversity in natural environments or wilderness. The range of conservation extends from the protection of the populations of specific species to the preservation of the entire ecosystem. For the protection of the specific species, within the wilderness, the protection of the entire wilderness itself is mandatory. The associated natural and cultural resource conservation are taken care of in the protected areas. This includes national parks, wildlife sanctuaries and protected and managed by employing legal measures and/or by other effective means.

Conserving natural biodiversity, including, geo-morphological, ecological, and biological features, ecological processes, and the different social, cultural, ethical, and aesthetic values, is the primary aim of the protected areas. It constitutes an internationally recognized management approach, adopted in specific areas, set aside for the purpose. The IUCN has developed methodologies which enable protection of biodiversity at the local, national, sub-national, and international levels. Local and national governments support and govern these protected areas and, sometimes, even private organizations and even individuals, who are very conscious of the importance of maintaining biodiversity on planet Earth, support such initiatives. Nomenclature, system of governance, approach to management etc., which related to the protected areas vary from place to place and also, from nation to nation.

Inasmuch as management aspects are concerned, while the management approaches might differ, the objectives of management must be well defined and fully met with. The IUCN has developed a system, recognized by prominent international organizations, such as the United Nations, and several national governments, for meeting the management objectives, by categorizing the protected areas to fully meet with the management objectives. Even the Convention on Biological Diversity (COD), ratified at the Earth Summit in 1992, in its Goal 43, Paragraph 4.3.7 of the Program of Work on Protected Areas, calls for using the IUCN system for management of protected areas. The IUCN management practices have become a worldwide standard for the planning, establishment, and management of protected areas. Table 5.2 summarizes the categories and their salient features.

There are, as many as 202,467 protected areas throughout the Planet Earth as on October 2017 according to the World Database on Protected Areas (WDPA). Protected Areas on Planet Earth cover about 200,000,000 km^2 or 14.7% of planet Earth's geographical area, excluding Antarctica. The Convention on Biodiversity under the Aichi Biodiversity Targets has set 17% area to be covered for biodiversity conservation in 2020. Hence, there is still a gap of 2.3% to achieve the stipulated target of protected areas. Part from the land area, about 10% of the territorial waters, are also are protected.

It is not that all the countries on Planet Earth have an equal proportion of land under protected area. Protected Areas in counties, such as, Afghanistan, Libya, Turkey, Lesotho, Mauritania, Syria, Yemen, Somalia, Barbados, Bosnia and Herzegovina, are almost negligible, ranging from 0.10% in Afghanistan to 1.40% in Bosnia and Herzegovina, as per the statistics of 2018. On the other hand, some countries, notably New Caledonia, Venezuela, Slovenia, Bhutan, Brunei, Seychelles, Hong Kong SAR, China, Greenland, Luxemburg and Congo have a very large proportion of their lands under protected areas, ranging from 54.40% in New Caledonia to 40.74% in Congo, as per the available statistics of 2018. Ten countries with minimum and 10 counties

Table 5.2 The IUCN categories of protected area management and their brief description

IUCN category	Description
Ia—Strict nature reserve	Strictly protected areas set side to protect biodiversity, including, as far as possible, the geological and geomorphological features; human interference is strictly controlled
Ib—Wilderness area	Large unmodified or slightly modified areas retaining their natural features; no permanent or significant human habitation
II—National park	Large natural or near natural areas set aside to protect large-scale ecological processes species and ecosystem characteristics; provision of educational, scientific, spiritual and recreational opportunities
III—National monument or feature	Areas set aside for protecting a specific national monument, for example, a geological feature like a grove, a landform, sea mount etc.
IV—Habitat/ species management area	Emphasis will be on the protection of specific species or habitats
V—Protected landscape/ seascape	Protection of the areas of distinctive characteristics, for example, significant ecological, biological, cultural and scenic value), created due to people's interactions or specific natural factors, as well as integrity of such interaction
VI—Protected area with sustainable use	Large area, mostly maintaining of Natural Resources natural conditions, with a proportion under sustainable natural resource management set aside for protection and biodiversity conservation of natural resources

Source Compiled from https://www.biodiversitya-z.org/content/iucn-protected-area-management-categories.pdf

with maximum percentage of the protected areas are graphically shown in Figs. 5.1 and 5.2.

A protected area under the in situ conservation strategy is ecologically more stable, healthier, vibrant, and resilient than an anthropogenic one managed in ex situ conditions. A pivotal role is played by the protected ecosystems which help build and maintain its own microclimate. Some of the vital attributes are the following:

1. Native species and their sub species are conserved and maintained
2. The size of the communities is maintained
3. Conservation of the genetic diversity of all species involved
4. Alien species invasion is prevented, which ensures non invasion of native ones

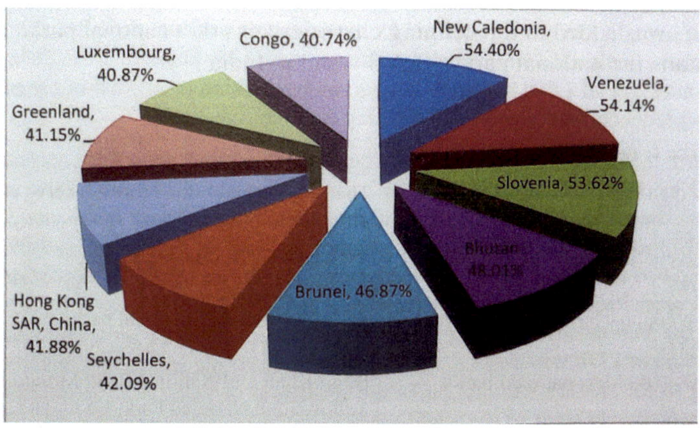

Fig. 5.1 Ten countries with minimum (almost negligible) coverage of terrestrial protected areas. *Source* World Database on Protected Areas (WDPA)

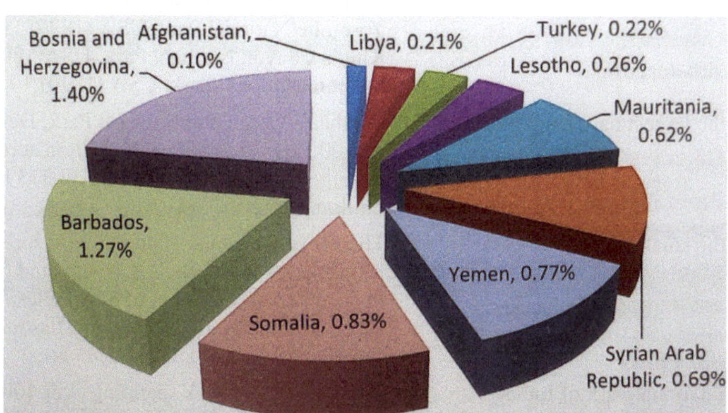

Fig. 5.2 Ten countries with maximum percentage of terrestrial protected areas. *Source* World Database on Protected Areas (WDPA)

5. Shifting species and their habitats in response to environmental changes is made possible.

National Parks

The welfare of wildlife is maintained in a national park. Agriculture, forestry, agro-forestry, grazing etc. is not allowed. A sovereign state owns a national park, which can own a reserve of natural, semi natural or developed area. People cannot enjoy private ownership rights. Hunting, exploitation of resources, and habitats manipulation are strictly prohibited. In 1969, IUCN did not fix the area size of a national park, and, apart from other conditions, only referred to it as "a relatively large area". In 1971, a minimum of 1000 ha "within zones in which protection of nature takes precedence" was added to the defining characteristics of a national park. Tables 5.3 and 5.4 details the national parks, globally and in India.

Table 5.3 The features of global national parks

Oldest protected area	The Tobago main ridge forest reserve established in 1776
Which is the continent having most of the national parks	Asia
Which is the country having most of the national parks	Australia with 285 national parks
Which are the 10 oldest national parks	Bogd Khan Uul National Park, Mongolia (1778) Yellowstone National Park, USA (1872) Royal National Park, Australia (1879) Banff National Park, Canada (1885) Glacier National Park, Canada (1886) Yoho National Park, Canada (1886) Tongarino National Park, New Zealand (1887) Yosemite National Park, USA (1890) King's Canyon National Park, USA(1890) Sequoia National Park, USA (1890)
Which are the largest national parks	Northeast Greenland National Park, Denmark, 972,000 km^2 (more than 24 million acres); Wrangell—St. Elias in Alaska, 32,375 km^2 (8 million acres)
Which are the smallest national parks?	The Moyenne Island National Park spread over just 23 acres in Seychelles, designated National Park by the Government of Seychelles, in 2008; Gateway Arch National Park in Missouri, USA spread over 0.7804 km^2 (192.83 aces)
The biggest national park of India	Hemis National Park spanning over 4000 km^2
The oldest national park of India	Jim Corbett National Park (1936)

Table 5.4 The ten largest national parks in India

Rank No.	Name	State	Area (km^2)
1	Hemis National Park	Jammu and Kashmir	4400.0
2	Desert National Park	Rajasthan	3162.0
3	Gangotri National Park	Uttarakhand	2390.0
4	Namdapha National Park	Arunachal Pradesh	1985.2
5	Khangchendzonga National Park	Sikkim	1784.0
6	Guru Ghasidas (Sanjay) National Park	Chhattisgarh	1440.7
7	Gir Forest National Park	Gujarat	1412.0
8	Sundarbans National Park	West Bengal	1330.1
9	Jim Corbett National Park	Uttarakhand	1318.5
10	Indravati National Park	Chhattisgarh	1258.4

Wild Life Sanctuaries

When a specific area of land is legally protected or reserved only for the conservation of wild animals, it is known as a "sanctuary". Habitat of wild animals and their surroundings are given due protection for their safety, life in comfort, free movement and breeding. Poaching, capturing and killing of animals in a sanctuary are very strictly prohibited. There are some anthropogenic activities, such as, harvesting of minor forest products and timber. Private ownership rights may be given provided wildlife is not affected adversely. Details of some of the planet Earth's biggest wildlife sanctuaries which give protection to wild animals in terrestrial and marine habitats are presented in Table 5.5. Natural Park of the Coral Sea encompassing an area of 1,294,994 km^2 and Pacific Remote Islands Marine National Monument with an area of 1,269,094 km^2 are among the world's largest animal wildlife sanctuaries. In India, wildlife sanctuaries are classified as IUCN Category 1V, protected areas. There are as many as 543 wildlife sanctuaries in India, covering a total area of 118,918 km^2. Among these sanctuaries, 50 are governed by the Project Tiger, aimed at conserving the famous Bengal tiger.

Table 5.6 provides details of the important wildlife sanctuaries in India. India hosts and provides sanctuary to a number of endangered animals. It befits the country's ethos.

What is a Biosphere Reserve?

It was in 1975, as part of the United Nations Educational, Scientific and Cultural (UNESCO) program, that the concept, "Man And Biosphere" (MAB), evolved, which has led to the creation of a special category of "Protected Areas"—land or coastal environments—with mankind as a central and integral component of the

Table 5.5 Some of planet Earth's largest wildlife sanctuaries

Name of sanctuary	Where located	Remarks
Coral Sea National Park Area:1,294,994 km^2, established in 2014	Approximately 1200 km east of Australia encompassing New Caledonia, lagoons of New Caledonia and the New Caledonia Barrier-Reef—world's second largest Barrier Reef	Protects native wildlife, e.g. 25 species of marine mammals, including manatee-like dugongs, 48 shark species, 5 sea-turtle species, and 19 species of nesting birds, one of world's 7 natural wonders—the Great Barrier-Reef—is also given adequate protection
Kavango-Zambesi Transfrontier Conservation Area, 517,998 km^2 established in 2011	Protects Africa's largest contiguous population of elephants, cheetahs, hippopotamus, African wild dogs, more than 600 bird species	Around the convergence of the borders of Angola, Zambia, Namibia, Botswana, and Zimbabwe; covers Victoria Falls and the Okavango Delta; includes 35 national parks, forest reserves, games reserves, wildlife sanctuaries
North-East Greenland National Park; Area: 971,246 km^2; established in 1974	An international biosphere reserve, inaccessible, protects Arctic Wilderness	Extending across half of the Greenland; an International Biosphere Reserve as many as 77 times larger than the Yellowstone National Park
Pacific Remote Islands Marine National Monument Area: 1,269,094 km^2 Established 2009, expanded 2014	West of Hawaiian islands, managed by US Fish and Marine Service	Comprises islands, atolls; more than 20 Mammals (whales, Dolphins etc.) protected
Papahanaumokuakea Marine National Monument; Area: 36,200 km^2. Established in 2006, expanded in 2016	Encompasses the Hawaiian Islands	Largest marine protected area in the world provides refuge to 7000 species of animals including Green sea Turtle, an endangered Species and the Hawaiian monk seal Wildlife Refuge and Midway Atoll National Wildlife Refuge; comprising 10 islands and atolls of the North Western Hawaiian Islands, larger than all National Parks in USA, also of Germany, managed by National Oceanic and Atmospheric Administration (NOAA) and Interior's Fish and Wildlife Service (FWS) of USA

Table 5.6 Details of the important wild life sanctuaries of India

Where sanctuary is located	Area	Which animal/animals involved
Karakoram (Nubra Shyok) Wildlife Sanctuary, Ladakh	5000	Tibetan antelope, wild yak, leopard, Himalayan mouse lynx, tufted duck, booted eagle, golden eagle, grey plover
Wild Ass Wildlife Sanctuary, Gujarat	4953.71	Ass, nilgai bustard, dalmatian pelican, hawks, harriers, falcons, Indian gazelle, desert fox (Indian and white footed), Jackals, African lynx
Changthang Wildlife Sanctuary, Ladakh	4000	Wild yak, Tibetan wolf, brown bear, bharal mormot)
Nagarjuna Srisailam Wildlife Sanctuary, Telengana	3568	Tiger, panther, black bucks, mouse deer, sambhar, spotted deer, jackals, wild boar, wild bear, sloth bear, leopard, Indian giant squirrels, crocodile, python, cobra
Hastinapur Wildlife Sanctuary, Uttar Pradesh	2073	Ganga river dolphin, Swamp deer, smooth-Coated otter, short-toed Snake eagle, Egyptian vulture, white-eyed buzzard, Indian grey hornbill
Gir Wildlife Sanctuary, Gujarat	1153.42	Asiatic lion, Indian leopard, Striped hyena four-horned antelope, wild boar, Mugger crocodile, Indian cobra, tortoise, monitor-lizard
Chilka lake bird Sanctuary, Odisha	990	Water fowl, duck, crane, sand piper, golden plover, flamingo
Kedarnath Wildlife Sanctuary, Uttarakhand	975.2	Himalaya black bear, Indian jackal, red fox, Himalayan monal, snow partridge, grey-cheeked warbler, Himalayan pit viper
Annamalai Sacuary, Coimbatore, Tamil Nadu	958	Tiger, panther, elephant, spotted deer, sloth bear, sambhar, barking deer, wild dog
Manas Wildlife Sanctuary, Assam	950	Tiger, panther, wild boar, rhino wild buffalo, golden languor, swamp deer, wild dog
Bhimbandh Wildlife Sanctuary, Bihar	681.99	Tiger, leopard, wild bear, nilgai, cheetal barking deer, langoor, Indian hare, pea fowl, van murgi
National Chambal Wildlife Sanctuary, Uttar Pradesh	635	Ganga river dolphin, gharial, striped-hyena, Indian wolf, red-crowned roof turtle, mugger crocodile, smooth-coated otter
Achanakmar Wildlife Sanctuary, Chattisgarh	551.55	Indian leopard, sloth bear, wild boar, blackbuck, sambhar, nilgai, chital, striped-hyena, Indian jackal, four-horned antelope
Balaram Ambaji Wildlife Sanctuary, Gujarat	542.08	Indian leopard, sloth bear, nilgai, striped-hyena, Indian fox, Indian civet, Indian porcupine, Indian pangolin
Abohar Wildlife Sanctuary, Punjab	186.5	Black buck, wild boar, blue bull, porcupine black duck

(continued)

Table 5.6 (continued)

Where sanctuary is located	Area	Which animal/animals involved
Interview Island Wildlife Sanctuary, Andaman and Nicobar	133.87	Feral Islands elephants, Spotted-deer. wild pig, three-striped palm squirrel
Keoladeo Ghana Bird Sanctuary, Rajasthan	28.73	Siberian crane, stork, egret, heron, spoon bill, spotted-deer, black buck, wild boar, blue bull

Note Area in square kilometers

entire system. The following philosophical ideas emerge from the basic concept, focusing on conservation, utilization and socio-economic considerations, which have the following contours:

1. In-situ conservation of all living forms with the associated support system
2. Reconciling biodiversity, with its important parameter, its sustainable use
3. Learning areas for sustainable development under the umbrella of diverse ecological, socio-cultural and economic contexts
4. Establishing sites for testing inter-disciplinary approaches/methodologies to sustainable management of available natural resources
5. Ecosystem management provides an opportunity for the participation by local populace in sound resource management
6. Sharing of the benefits of sustainable management systems with all the people involved, equally and equitably
7. To establish ecological spaces to resolve conflicting interests between the public system and the local people involved in an environment, relating to sustainable management of biodiversity resources, especially with regard to conservation and utilization
8. Providing local solutions in the context of global problems.

Currently, there are 701 biosphere reserves, globally, within the geographical (territorial) boundaries of 124 countries. There are 195 countries in the world, of these, 193 are sovereign member states of the United Nations, world's largest inter-governmental body, and only two are not members: a. The Holy See (Vatican in Rome, Italy) and the Palestine State. These 124 countries include 21 transboundary sites which belong to the World Network of Biosphere Reserves, an organization which promotes international cooperation by employing and exchanging experience and technical know-how, capacity building, and promotion of the "Best Practices" for the sustainable management of these biosphere sites.

Local communities, including their stake holders, must be actively and keenly involved in, firstly, the planning, and later, the sustainable management of the biosphere sites. Their close interaction must ensure the following three objectives:

1. Providing the logistic support for education, research, training and subsequently effectively monitoring all the activities in a sustainable manner
2. Conservation of biodiversity and their cultural diversity

3. Ensuring socio-culturally and environmentally sustainable economic development.

The natural resource management in a sustainable manner, focusing on biodiversity conservation, further promoted by employing logistic support for education, research, training and monitoring, becomes the foundation for sustainable development—the economic development reconciled with the environment.

A biosphere reserve maintains the above mentioned three functions, effectively, based on the following three zones as noted below:

A. Core Zone
B. Buffer Zone
C. Transition Zone.

The description of the above three zones is as follows:

A. Core Zone: It is the core conservation area functioning to conserve biodiversity at the landscape, ecosystem, species, and, genetic levels. At the core zone, all human activities are strictly prohibited. Natural evolutionary processes prevail in this zone and this zone is thought to be ecologically vibrant, healthy, and sustainable. However, if a research project is interesting, and, vital in the overall interest of all living species involved, such a research project may be allowed to be carried out in the core zone.
B. Buffer Zone: The buffer zone surrounds the core zone and this, in turn, is surrounded by the transition zone. The buffer zone is sandwiched between the zone meant for conservation and the zone with human activities. This zone is often used for educational activities, scientific research, training and monitoring etc. This zone is also open for tourism, both national and/or international.
C. Transition Zone: The outermost area of the biosphere reserve comprises what is meant for human intervention to foster developmental activities for economic gains. The developmental activities encouraged must be the ones, which are ecologically sound, environmentally safe and, socio-culturally just and equitable.

Table 5.7 lists some of important biosphere reserves in India. Most of the biosphere reserved in India are protected by the rules and regulations of UNESCO. Five of them have been recognized as UNESCO Heritage Sites.

Table 5.8 details the different Hermitage Sites in India.

The Sacred Forests or Groves

The Sacred Forests or Groves represent an alternative approach to biodiversity conservation which involves the principles which arise from a diversity of prevalent cultures or social taboos, rather than approaches based on legality. Biodiversity protection and conservation in the sacred groves have been a tradition in most of the Asian continent. India is specially known for this age-old tradition. Sacred seeds, sacred trees,

Table 5.7 List of the important biosphere reserves in India

Name and where located	Area (km^2)	Important animals
Great Rann of Kutch in Gujarat	12,454	Indian Wild Ass
Gulf of Mannar Marine National Park in Tamil Nadu	10,500	Dugong
Sunderban Biosphere in West Bengal	9630	Royal Bengal Tiger
Cold Desert in Himachal Pradesh	7770	Snow Leopard
Nanda Devi Biosphere Reserve in Uttarakhand	5860	Snow leopard, Himalayan Black Bear
Nilgiri Biosphere Reserve in Tamil Nadu, Kerala and Karnataka	5520	Lion-tailed macaque, Tiger, Nilgiri Tahr
Dihang Dibang in Arunachal Pradesh	5112	Mishmi Takin, musk deer
Seshachalam Hills in in Andhra Pradesh	4775	Slender loris
Simplipal Biosphere Reserve in Odisha	4374	Gaur, royal Bengal Tiger, Asian elephant
Pachmarhi Biosphere Reserve in Madhya Pradesh	4189.78	Giant squirrel, flying squirrel
Achanakmar-Amarkantak Biosphere Reserve in Madhya Pradesh and Chattisgarh	3835	Four-horned antelope, Indian wild dog, sarus crane, white-rumped vulture, sacred-grove bush frog
Agastyanalai Biosphere Reserve in Kerala and Tamil Nadu	3500	Nilgiri tahr Asian elephant
Manas Biosphere Reserve in Assam	2837	Asian elephant, tiger, Assam roofed turtle, hispid hare, golden langur pygmy hog
Khangchendzonga Biosphere Reserve in Sikkim	2620	Snow Leopard, Red Panda
Great Nicobar Biosphere Reserve	885	Saltwater Crocodile
Nokrek Biosphere Reserve in Meghalaya	820	Red Panda
Dibru-Saikhowa Biospere Reserve in Assam State	765	White-winged wood duck, water buffalo, black-breasted parrot bill, tiger capped langur
Panna Biosphere Reserve in Madhya Pradesh	298.98	Tiger, chital, Chinkara, Sambhar, Sloth bear

Table 5.8 Details of the UNESCCO recognized heritage sites of India

S. No.	Protected area (UNESCO heritage site)	In which state located	Area in square km
1	Kaziranga National Park	Assam	430
2	Keoladeo Ghana National Park	Rajasthan	29
3	Manas Wildlife Sanctuary	Assam	950
4	Sunderbans National Park	West Bengal	2585
5	Khangechendzonga National Park	Sikkim	1754

sacred forests, sacred rivers, sacred lakes and so on, underly the basic sanctity of these materials to Indian life. It is very interesting to note that almost every ancient temple in India has a sacred tree, mostly Peepal/Banyan, in front of it, which is never pruned and it's seeds are used as breeder seeds. Its fruits and seeds are dispersed to other places for multiplication. And the tree is never cut, in fact, it is irreligious, as per Hindu beliefs, to cut a banyan tree in front of a temple.

The religious sanctity for some specific forests is dearly upheld by the majority religion of Hindus in India. There is, but, minimum human intervention in sacred groves/forests. Both logging and poaching are strictly prohibited in these sacred groves/forests. Hence, there will only be minimal ecological degradation in these community controlled sacred groves/forests. As a result of this, such community protected forests, in a sense, represent pristine forests. These forests are healthy and vibrant and serve as home to native bio diversity. Many sacred groves/forests are surrounded by degraded landscapes. In such a situation, the sacred groves/forests serve as oases of greenery, with native species flourishing in the natural environment.

The cultural ethos of India is, indeed, very diverse and very tolerant. The maintenance of the sacred groves/forests is so very symbolic of this culture. In India, the number of sacred forests could be well over 100,000, according to an investigation carried out by Ormsby et al. in 2010. This approach of in situ conservation has greater acceptance by the local community. Many states in India, notably, Himachal Pradesh, Madhya Pradesh, Bihar, Maharashtra, Manipur, Meghalaya, Kerala, Karnataka, and Tamil Nadu, have many sacred forests being managed by local communities. Many are, indeed, endangered, but, endemic species are prospering in these people/community-protected ecosystems.

The sacred groves/forests in India are known by various local/regional names based on their geographical locations and cultural zones. For instance, in both Madhya Pradesh and Himachal Pradesh, they are referred to as "***Dev Van***" (Forest of God), and, in other states, the community protected sacred forests are locally known as ***Sarnas*** in the State of Bihar, ***Oran*** in the State of Rajasthan, ***Devarakadu*** in the State of Karnataka, ***Kavu*** in the State of Kerala, ***Devrai*** or ***Dev Van*** in the State of Maharashtra, ***Lai Umang*** in the State of Manipur, ***Law Kyatang*** or ***Law Lyngdoh*** in the State of Meghalaya ***Sarpa Kavu*** (**meaning forest of the snakes**) in the State of Tamil Nadu. Cultural rituals may differ from one state to another, or geographical location, but, the ways and means by which the sacred groves/forests are protected are almost similar. Again, the species of plants/trees considered sacred might also vary from state to state and also depending on the geographical area where they are located, but, this, in reality is very vital for the maintenance/conservation of biodiversity.

In the ancient Hindu tradition of India, celebrating biodiversity is a way of life and the ethos of the country. In fact, a number of festivals are organized in India which are devoted to specific species of plants. Every Hindu festival is associated with a specific plant species and/or food/feed, derivable from a variety of plant species. The main trees celebrated in India are the following:

1. *Azadiracta indica* (Neem tree, Margosa)

2. *Ficus bengalensis* (Banyan tree)
3. *Ficus religiousa* (Peepal tee)
4. *Aegle marmelos* (Bilwa or Bael a medicinal plant)
5. *Musa paradisiaca* (Banana).

These trees are celebrated on different occasions in India (Anubhav et al., 2009).

Every Hindu family has a plant of specific spices for celebration on important religious occasions. For instance, Holy Basil popularly known as *Tulsi/Tulasi* (*Ocimum sanctum and Ocimum tenuiflorum*)—are planted at home, in the front, facing east. Every evening, at dusk, (sunset), an oil lamp is lit and placed before the *Tulsi* plant, which is an important religious practice. There are numerous other plants, trees, shrubs, and herbs, which are given due protection/conservation as they also contain ingredients of great medicinal value. Pseudo cereals, amaranth and buckwheat are commonly used for eating on the days when Hindus fast. Sacred groves are not only sacred ecosystems, functioning as a very rich repository of nature's unique biodiversity, but, also, as a product of the socio-ecological-cultural philosophy our forefathers have been cherishing, since olden days. Not only are they very valuable biodiversity, but, soil, and water, as well, are conserved by these ecosystems, and, they act also as agents regulating climate change, which has become a modern day threat, which is very vital for life to blossom and thrive on planet Earth. This kind of informal protection takes care of the rare, endangered vulnerable species of both plant and animal species, in their natural habitats, which can, more often than not, be missed through the use of formal protection measures applicable to these species.

The Sacred Lakes

The different Indian communities also protect many water bodies. An illustrious example of this is the *Khecheopalri* Lake in Sikkim, which has been declared a sacred lake by the local community. The local community ensures/maintains not only a good physical state of the lake, but also, its sanctity by preventing any pollution made by man or animal. The Sacred Lake protects the local fauna and flora of the local aquatic systems.

Like the above, in Gujarat State, in India, *Bindu Sarovar* and *Narayan Sarovar*, in Kerala State, the *Pampa* River, adjacent to the famous *Ayyappa* Temple, in the State of Rajasthan, the *Pushkar Lake*, and the most famous *Manas Sarovar* or *Mansarovar*, on way to the Kailash mountains (which is in Tibet, unfortunately now under the control of the Chinese), are other important sacred lakes. Tibet is also known for its sacred lakes, such as, Lake *Namtso* (also called the Heavenly Lake), and *Basum* Lake. Also, *Tangra Yumco*, *Lake Rakas Tal* (also called the Devil Lake), *Lamu Nacuo* (also known as Goddess Skull Lake), are in Tibet.

It is the deep belief in religious sanctity that ensures and maintains the sacredness of the sacred lakes, kept free of external pollution, which serve as a healthy habitat for aquatic life. Sacred groves and sacred lakes both significantly contribute to protect and conserve the local biodiversity, with the overwhelmingly enthusiastic participation of the concerned local communities.

The Ex-situ Conservation

For the Nature's biodiversity, man is it's custodian. Continuous and greedy anthropogenic activities have led to the destruction of many species and their sub species in the natural/original habitat. Hence, conservation of biodiversity, at places away from their natural habitat, or ex. situ conservation is becoming increasingly important. Or critically important, so that the specific species is not lost forever. The establishment of germplasm banks, is the principal strategy of ex. situ conservation. The gene banks include botanical gardens, zoos, genetic resource centers, seedlings, seed, tissue culture and DNA banks.

The Gene Bank

A gene bank is a biorepository, where seeds and plant cuttings or vegetative part, from a large variety of plants are stored at controlled temperature, humidity and light. The principal role the gene banks play is the preservation of a specific genetic material. And, it is done in the following manner:

1. Through the creation of artificial ecosystems, such as ambient temperature, humidity and light conditions (to simulate photoperiods)
2. The conditions are artificially controlled
3. The specific genetic material is stored in appropriate nutrient media
4. The genetic material is frozen at very cold temperatures to – 96 °C, in liquid nitrogen.

The above procedures apply only to vegetative plant material. When it comes to preserving the specific genetic material from animal sources, the sperms and eggs are frozen in zoological freezers. For corals, fragments are taken and stored in water tanks under controlled ambient conditions, for temperature, humidity and light conditions (photoperiod). The genetic material preserved/conserved in a gene bank is meant to be made available for future use in research in plant and animal breeding.

The Technique of Cryopreservation

The word cryopreservation originates from the Greek word "Kruos" meaning, icy cold or frosty. It is a process of storage/preservation of biological samples (both of plant or animal origin) in sub-zero temperatures, at – 16 °C, using liquid nitrogen.

Temperatures to this very low level help maintain the survivability of the biological sample, with the thawing of the physiological conditions. Cooling at extremely low temperature, as mentioned above, causes the formation of crystals, within the cells leading to the rupture of cell membrane, which leads to the death of the cell. Biological material, including organelles, cells, tissues, extra cellular matrix, and, organ, are preserved by employing the cryopreservation technique.

The Botanical Gardens

A place dedicated to the conservation of biodiversity, especially if it refers to plant species, is called a botanic garden or botanical garden. A botanic or botanical garden collects, cultivates, preserves, and displays a wide range/variety of plants. This is a garden, especially of living plants, designed chiefly to illustrate the relationships within groups of plants, for scientific investigations, research and, ex-situ conservation of plant biodiversity. The collection of plants in the botanic garden will include specialized plants, for instance, cacti and other succulent plants, ornamental herbs and ornamental shrubs or plants, obtained from a specific geographical region, within a country, or, the planet Earth, at large.

The name "Arboretum" is used in the place of Botanic Garden, when the collection of plants refers to only trees. An arboretum is often associated with an animal or animals zoo or wildlife. The botanical gardens and arboreta are often managed by scientific research institutes or universities, especially, agricultural universities. In some cases, they are independent institutes in themselves and have their own agenda of research, education and conservation. There are more than 2000 botanical gardens and arboreta across the globe, which house more than 80,000 species of plants in ex-situ conservation. Many botanic gardens have their seed banks/vaults, tissue culture facilities, as well as technologies worth promoting ex-situ conservation.

The following are some the world's most prominent botanic gardens:

1. London Kew Garden, Britain
2. Longwood gardens, USA
3. Jardin Botanique Montreal, Canada
4. Hawaii Tropical Botanic Garden
5. Orto Botanico di Padova, Italy
6. Botanischer Garten Munchen, Germany
7. Singapore Botanical Gardens
8. Sydney Royal Botanical Garden, Australia
9. New York Botanical Garden, USA.

It was in 1544 that the Orto Botanico di Pisa, in Italy, was established which has the distinction of being the oldest botanical garden in the world. This botanical garden was run by the University of Pisa in Italy. Founded in 1545 and operated by the University of Padua, the Orto Botanico di Padua in Italy is sometimes regarded as the world's oldest academic botanical garden, because, it was not moved from the original place where it was established. However, it is the London Kew Garden, in Britain, which has the credit of having the most, 50,000, living plant species. In Asia, the largest botanic garden is the Gyeongsangbuk-do Arboretum, in South Korea, founded in 2001, which is spread over some 3222 ha, containing 179,226 living plants which belong to 1510 different species.

As far as India is concerned, the history of botanic gardens begins with the establishment of the Indian Botanical Garden or Royal Botanic Garden or Calcutta (the name now of Calcutta is Kolkata) Botanic Garden, established in 1786. It is spread over an area of 109 ha or 270 acres. The Indian Botanical Garden was later renamed as Acharya Jagdish Chandra Bose Indian Botanic Garden, in memory of the great Indian scientist Jagdish Bose. This is the largest botanical garden in India. The main attraction in this botanical garden is that it has the world's largest Banyan tree (*Ficus benghalensis*).

The Animal Zoos

A common strategy to support conservation of wild animals is to install ex-situ conservation through the establishment of animal zoos. These animal zoos are situated far away from the wilderness, often, within or adjacent to human habitation. And, they are open to visitors and tourists. A variety of animals have been protected by such measures. Zoos are also places of human recreation. However, wild animals in zoos often suffer stress and these ex-situ conservation fails to provide an environment conducive to the natural conservation and growth of wild animal protection. Creative breeding programs run in professionally managed zoos often do not succeed owing to the absence of the natural environment for animals to live and thrive.

In the world, there are about 800 professionally managed zoos, accommodating more than 3000 species of mammals, birds, reptiles, amphibians etc. Many of the animal species being conserved in the zoos include the rare and endangered species, and, also, those threatened by extinction or those on the verge of extinction.

The oldest zoo in the world is the Vienna Zoo, Tiergarten Schonbrunn (Animal Garden) founded in 1752 in Vienna, Austria, and continues to thrive successfully to date. The other important and often discussed zoos conserving a variety of fauna are the National Zoological Gardens of Africa, London Zoo, San Diego Zoo, Moscow Zoo, Toronto Zoo, Columbus Zoo and Aquarium, The Bronx Zoo, Berlin Zoological Garden in Germany, Dallas Zoo, Philadelphia Zoo, Oakland Zoo, Detroit Zoo, Los Angeles Zoo etc. in the USA.

What About the "Hotspots" of Biodiversity?

Planet Earth houses biodiversity, possibly, the only home in the universe for the same, known so far. The distribution of biodiversity on Planet Earth, is extremely non uniform. It is obvious, as biodiversity is an attribute of a complex of environmental factors. Some regions of planet Earth harbor a large number of both plant and animal species. The regions prospering with a very high degree of biodiversity are known as a "Megadiversity Zones". About twelve countries with 60–70% of the world's total biodiversity have been identified as megadiversity countries. India is one of the megadiversity countries. India constitutes to about 8% of the total global biodiversity despite occupying only 2.4% of the world's geographical area.

Norman Myer (1934–2019) elucidated the concept of the biodiversity "Hotspots" to designate priority areas for In-situ conservation and then elaborated the same in 1990 and 2000 (Myers, 1988, 1990; Myers et al., 2000). The hotspots of biodiversity are defined as the richest and the most threatened reservoirs of plant and animal life on planet Earth. The determination of the hotspot category is based on the following two important criteria:

1. The number of endemic species, that is, the species which ae not found elsewhere or in other areas: A hotspot of biodiversity must contain at the least, 1500 species of vascular plants (> 0.5% of the world's total as endemics.
2. The degree of threat to biodiversity is measured in terms of habitat loss: A hotspot should have at least 70% of its original natural vegetation.

On planet Earth, there are three dozen biodiversity hotspots which strictly fulfill the criteria stipulated, above. And, these are hotspots, where the conservation is in-situ. These hotspots represent just about 2–3% of the planet Earth's land surface, but, support, approximately, 60% of the planet Earth's plant species and 42% of the animal species—terrestrial vertebrate species, which are, the mammals, birds, reptiles and amphibian species. Many of the biodiversity hotspots exceed the above mentioned two criteria. For instance, the Tropical Andes in South America and Sundaland in Southeast Asia, the two biodiversity hotspots, have approximately 15,000 endemic plant species. Almost 95% loss of vegetation has been observed in some of the hotspots.

Which Are the "Megadiverse" Countries?

The following seventeen countries are classified as "megadiverse" countries:

The "Megadiverse" countries in South/Latin American continent are:

1. Brazil
2. Columbia
3. Ecuador
4. Mexico

5. Peru
6. Venezuela.

The "Megadiverse" country in North America:

1. The United States of America

The "Megadiverse" countries on the African continent:

1. The Republic of Congo
2. Madagascar
3. South Africa.

The "Megadiverse" countries on the Asian continent:

1. India
2. China
3. Malaysia
4. Philippines
5. Indonesia
6. Papua New Guinea
7. Australia.

The megadiverse countries represent more than two-thirds of all the known life forms. Most of the tropical rainforests and coral reefs are also included in these megadiverse countries. The megadiverse countries are spread in less than 10% of the global land surface, but, support more than 70% of the global biodiversity. Table 5.9 details the biodiversity hotspots, falling in various biogeographic regions.

A year following the publication of the work on biodiversity of hotspots by Norman Myers, Conservation International, an environmental organization, adopted the idea of protecting the biogeographical regions, exceptionally rich in biodiversity, but, threatened by various factors of destruction.

Out of the total, five of the world's biodiversity hotspots are located in India. There is enormous endemism in these hotspots and they accommodate a very large number of diverse flowering plants, mammals, reptiles, amphibians, and swallow-tailed butterflies. As the geo-biological regions know no international boundaries, the biodiversity hotspots in India extend to the boundaries of many of the neighboring countries and which are located in the Indian Ocean. For instance, such extensions can be between Indian and Sri Lanka, India and Nepal etc. They are described below:

The Five Global Biodiversity Hotspots:

1. The Himalayas: The entire Indian Himalayan region extending up to the parts of Pakistan, Tibet, Nepal, Bhutan, Myanmar.
2. Indo-Burma: Entire North Eastern India (except the state of Assam and Indian Union Territory of Andaman Islands) extended to Myanmar, Thailand, Vietnam, Cambodia, Laos, and southern China.
3. Eastern Himalayas: Part of the two hotspots, namely, Indo-Burma and newly distinguished Himalayas. Includes North-East India extended to Bhutan and

Table 5.9 Details of the global biodiversity hotspots

The biogeographic region	The biodiversity hotspots
South America	Tropical Andes, Atlantic Forest, Tumbes-Choco-Magdalena, Cerrado Chilean Winter-Rainfall-Valdivian Forests
The Caribbean	Caribbean Islands
North and Central America	Mesoamerica, California Floristic Province, Madrean Pine-Oak Woodlands, North American Coastal Plain
Africa	Madagascar and the Indian Ocean Islands, Coastal Forests of Eastern Africa, Guinean Forests of West Africa, Cape Floristic Region, Succulent Karoo, Maputaland-Pondoland-Albany, Eastern Afromontane, Horn of Africa
Europe	Mediterranean Basin
West Asia	Caucasus, Irano-Anatolian
East Asia	Mountains of Southwest China, Japan
South Asia	Indo-Burma, India, Myanmar, Western Ghats Sri Lanka, Eastern Himalayas
Central Asia	Mountains of Central Asia
South East Asia and Asia Pacific	Sundaland and Nicobar Islands of India, Wallacea, Philippines, South-West Australia, New Caledonia, New Zealand, Polynesia-Micronesia, East Melanesian Islands, Eastern-Australian Temperate Forests

Source Mittermeir et al. (1999). Hotspots: Earth's Biologically Richest and Most Endangered Terrestrial Ecoregions. Monterrey, Mexico: Cemex, Conservation International and Agrupacion Sierra Madre

southern, central, and eastern Nepal. Originally part of the Indo-Burma biodiversity hotspot. Numerous primitive angiosperm families, such as, *Magnoliaceae* and *Winteraceae*, and, primitive genera of plants, for example, *Betula* and *Magnolia* are seen to exist.

4. Western Ghats: Entire Western Ghats parallel to the Western Coast of the Indian Peninsula for about 1600 km, covering part of the Indian states of Karnataka, Maharashtra, Kerala and Tamil Nadu, with the two main biodiversity centers (a): the *Agastyamalai* hills in Tamil Nadu and Silent Valley in Kerala and (b): the *Amabalam* Reserve, extended up to Sri Lanka.
5. Sundaland: Nicobar Group of Islands in the Indian Union Territory, extended up to Singapore, Malaysia, Philippines, Indonesia and Brunei.

What Are the Efforts for Biodiversity Conservation Done Internationally?

The remarkable book by Rachel Carson, Silent Spring, in 1962 was the turning point in a global movement concerning the preservation of the environment. The Earth Summit, held in Rio di Janeiro, Brazil, in 1992, has been a milestone in the history of the environmental movement. The first, ever, conference on Environment and Development was held in Stockholm, Sweden, in 1972. The concept of Sustainable Development laid down by the Brundtland Commission Report "Our Common Future", in 1987, called for reconciliation of environment and development. The Earth Summit was something extraordinary in the sense, it was the first to raise biodiversity-related issues on an international platform. The Biodiversity Convention resulted from the Earth Summit came into force on 29 December 1993. The Biodiversity Convention had the following three key objectives:

1. Conservation of Biodiversity
2. Sustainable Use of Biodiversity
3. Fair and equitable sharing of the benefits arising out of the utilization of genetic resources.

Many projects on biodiversity conservation, and, the appropriate development of biosphere reserve, national parks, and wildlife sanctuaries are being supported by the international organizations, such as, the International Union for Conservation of Nature (IUCN) and World Wildlife Fund for Nature (WWF). Projects relating to ex-situ and on farm biodiversity conservation in agriculture, forestry, and livestock sectors are supported by the Food and Agriculture Organization (FAO) in Rome. Numerous non governmental organizations, all over the world, are also engaged in biodiversity protection, conservation and amelioration.

In the effort/endeavor to conserve the global biodiversity hotspots, several international organizations are involved. However, only a small proportion of the land area falling within the defined biodiversity hotspots is currently protected. The following are some of the organizations committed to the protection and conservation of biodiversity hotspots, including biodiversity within these hotspots, globally:

1. The National Geographic Society
2. Critical Ecosystem Partnership Fund (CEPF)
3. Word Wide Fund For Nature (Global 200 Ecoregions)
4. Plant Life International
5. Birdlife International
6. Alliance for Zero Extinction.

Biodiversity and Sustainability

In the words "biodiversity parlance", sustainability does not denote a fixed/rigid state of things. It is, indeed, a very dynamic phenomenon. A phenomenon that tends to advance toward a higher state of "being". The sweep/range of a journey in sustainability stretches from nothing to infinity. From a simple system to a lot more complex system, from an unstable or less stable system to a more stable system. If ones uses the ecological language, the transformation of an unproductive desert land into a forest in ecological climax, over a long stretch of time, is how the phenomenon of sustainability operates.

It was during the World Conservation Strategy in 1980 (IUCN-UNEP-WWW 1980) that the term "sustainable development" was first used. However, this term was brought under limelight in the world, for the first time, during the United Nations World Commission on Environment and Development (WCED) Report, popularly known as the Brundtland Commission Report titled "**Our Common Future**", in 1987. Sustainable Development in the WCED Report has been defined as follows:

> Development that meets the needs of the present without compromising the ability of the future generations to meet their own needs.

In spirit and principle, this indeed, is a very noble, lofty, thought. However, in practice, this aspiration is far from met. In this author's well-considered opinion, it is the "anthropogenic greed" of man, that has come in the way of implementing/fulfilling this noble/lofty thought. How else, can we justify the plunder of the Earth, and, more recently even the bottom of the oceans, as discussed in the previous sections of the chapters of this book, so very blatantly?

Sustainable development, hence, views the socio-economic well being through the resolution of environmental issues. The word sustainability, often, is employed as synonymous with sustainable development, that generally refers to the human ability to live constantly in tune with the capacities of the biosphere, and hence, biodiversity is the crux for building up the biosphere's capacities.

There are many an instance, during the journey of man, from being a hunter and gather, to tending the plants in a fixed area, that lead to the most debatable agricultural revolution. This led to paying a huge price on biodiversity, be it plant, animal or even soil biodiversity. And, nowhere in the history of mankind, such a huge compromise has been made as in the case of so-called "green revolution", of recent memory, wrought in. This author has termed this as "anthropogenic greed".

A Vital Dimension of Sustainability

The ecological sustainability is a vital dimension of sustainability, which, more often than not, is distorted in favor of the actors involved. What history has taught us, is that, more often than not, it is delinked from its deeper meaning, which clearly spells out its ecological dimension and meaning, and is most often targeted for economic benefits. For example, the so-called highly chemical-centric and soil extractive farming, euphemistically called the "green revolution" is the most illustrative example of this distortion (Nair, 2023a, 2023b). Of course, sustainability may have many dimensions, including, socio-economic sustainability. However, no dimension, other than the ecological dimension, is independent. All dimensions, eventually, are expressed via the ecological dimension. For instance, Nair (2019a, 2019b, 2019c) has clearly argued that what the green revolution has done to world soil resources is a most intransigent exploitation, leaving the soil resources of south Asia, in particular of India, totally degraded, thus, severely and very adversely affecting their ecological sustainability/productivity. There can be no sustainability without ecological sustainability (Singh, 2019, 2020).

Without the considerations of ecosystems a sustainable economy cannot measured. All kinds of livelihood and economic dimensions ultimately survive and function only through the efficient functioning of the ecosystem. A vibrant and sustainable economy can only be ensured through a healthy and vibrant ecosystem. And, a healthy and vibrant ecosystem depends on the biodiversity it provides and nourishes. Biodiversity on the other hand, structures and maintains an ecosystem. In essence, biodiversity is the basis of environmental sustainability, which in turn, serves as the basis of other dimensions of sustainability.

How Does Biodiversity Connect with Sustainability?

Sustainability is maintained by biodiversity through its action of induction, building it up, and enhancement. The Nature's resilience is enhanced by biodiversity, and, this resilience is the capacity of Nature to recuperate from physical and biological shocks (Singh, 2019, 2020). The higher the level of biodiversity in Nature, the higher the degree of resilience. In natural ecosystems, sustainability is attained through resilience. The higher the degree of resilience, the higher the state of sustainability.

The predators, both herbivores and carnivores, are provided with a wide range of choices, as preys, by the biodiversity, to feed on and survive in an ecosystem. For instance, different insects feed on different types of plants. If there is a high level of plant diversity, in an ecosystem, they would all survive and flourish, without any considerable pressure on defoliation, as happens frequently when some insects defoliate the entire plant. Monocultures are, more often than not, become victims by specific insect pests. One of the most illustrative and eye-opening examples of this, in the world, is the complete devastation of the rice crop in Asia, during the green revolution phase, by the brown plant hopper insect, *Nilaparvata lugens* Stahl. The insect attack makes the rice leaves turn brown and dry, leading to the entire death of the rice plant, called the "hopper burn". Also, monocultures introduce vulnerability in the ecosystem. Physical factors like strong winds, storms heavy rains, fires etc., and by biotic pressure, such as intensive grazing, insect attack and attack by plant pathogens, both fungus and bacteria, etc., also cause serious damage to monocultures.

Contrarily, heterocultures (heterogenous) ecosystems are, invariably, more resilient, compared to monocultures, and, consequently more sustainable.

Biodiversity attributes, translating into the ecosystem and socio-economic sustainability, are complementary to each other, one influences and is influenced by the others. Based on the natural links between biodiversity and sustainability, as discussed above, ecologically sustainable and vulnerable ecosystems can be distinguishable. The ecosystems featuring a higher level of biodiversity are more sustainable than the ones with lower levels of biodiversity. In other words, the ecosystems with a lower level of biodiversity are ecologically more vulnerable than the ones with a higher level of biodiversity. These traits are depicted in Table 5.10.

The planet Earth's biodiversity is, unquestionably, the basis for survival for all—mankind, animal, plant and soil. When mankind fails to realize the cruciality of the link between the above-mentioned three words, destruction and death are the only consequence. For instance, for long, soil has been considered inanimate, and, it was this author who, for the first time, coined the phrase, "soul of infinite life", for soil, during the 13th World Congress of Soil Science held in Hamburg, The Federal Republic of Germany, in August 1986. When soil dies, mankind dies, civilizations are swept off the face of planet Earth. When the North African soils desertified, the Roman civilization collapsed. Nearer home in South Asia, in India, in particular, there is another threat unfolding. The highly chemical—centric extractive farming, euphemistically called the "green revolution"—a source of excessive greenhouse gas emission due to unbridled use of nitrogenous fertilizers to prop up this farming system of monoculture, is leading to huge soil degradation in South Asia, especially in India, and, this in turn, will become a threat to food security. The phenomenon is a slow path of destruction of soil resources, which in centuries to come, unless checked, will lead to the extinction of the human population in this sub continent, primarily, India.

Table 5.10 Some very illustrative examples of ecologically sustainable and vulnerable ecosystems

Ecologically sustainable ecosystems	Ecologically vulnerable ecosystems
Natural forests at the ecological climax	Forests managed by mankind
Heterogenous forests	Monoculture forests
Oak-type forests in temperate regions	Apple orchards
Agroecosystems (forests-cultivated lands based farming systems)	Cultivated croplands
Agroforestry (cropping systems integrated with trees)	Cultivated croplands
Mixed farming systems	Monoculture-based farming systems, the most illustrative example being the highly soil-extractive green revolution

Biodiversity, Economic Progress and Sustainability—the three most closely inter-linked words in the history of mankind

The green revolution has also caused global warming. According to the calculations of this author (Nair, 2019a, 2019b, 2019c) it adds up to about 35% to global warming, because, the gaseous nitrous oxide, a byproduct of urea fertilizer hydrolysis in soil, stays put in the stratosphere capturing the radiant sunlight and adds to global warming. The recent heatwave in Europe is a dire warning to mankind of the impending danger. California's death Valley, often among the hottest places on planet Earth, is likely to register new peaks in heat, with the mercury possibly surpassing 54 °C. In Southern India, the state of Kerala, is experiencing the worst heat waves in July, when there should be incessant rainfall. At the time of writing this book, in the place of excessive rains, the ambient temperature is around 35 °C—a most unusual phenomenon.

Development, speaking in the words of the great President of America, Abraham Lincoln, is the economic well being of the people, by the people, for the people. Development is a socio-economic advancement, in essence. The process, inevitably, takes place through human interaction. But, the environmental resources should be thoughtfully and equitably used. Dependent on nature's diversities, development is a dynamic process. Changes brought about as a result of the development processes cost nature's diversities—ecosystems, species, genetic resources, and, the physical environment.

When development keeps pace with nature's regenerative capacities, in a harmonious manner, it can be termed as "sustainable development", but, on the other hand, if it exceeds nature's regenerative capacities, it is "unsustainable development". The classic example of this is the unsustainable chemically—centric, soil extractive "green revolution", which has left more than a third of India's geographical area with degraded soils, where not even a blade of grass will grow, without huge investments in soil reclamation.

Sustainable development is characterized by a larger proportion of resource size, the highest possible level of nature's biodiversity, conservation-oriented changes in resource allocation/utilization for various purposes, and, in resource management, environmentally safe technological processes, and social justice. In fact, equity must be the driving principle of resource allocation/utilization. "Anthropogenic Greed" must never be permitted to take the upper hand. Such a kind of development confirms to the Brundtland Commission's ethos. If the processes of development are resource damaging, biodiversity eroding, unwise, haphazard, and driven by what this author calls the "anthropogenic greed" not by "anthropogenic need", then environmental degradation and many other socio-cultural tensions would be the natural outcomes, and, such a resource-depleting development would, inarguably be, unsustainable.

The Contrasting Features of Natural and Anthropogenic Ecosystems

One must clearly understand that, there are functional differences between a natural ecosystem and an anthropogenic (man made, artificial, environmental changes brought about by man) ecosystem. It might sound unnatural when one uses the word "artificial" in the context of ecosystems. In reality, this is true. Most of the land on planet Earth bears, what one can refer to as "man made", artificial or anthropogenic ecosystems, for, there has been enormous human intervention in Nature, not allowing the ecosystems to evolve the way they could. With continuous, unabated, and ruthless exploitation of the resources of the planet Earth's ecosystems, consequently, aimed at extracting the maximum benefits from natural ecosystems, most of them are turning into far less functional entities of Nature, than they ought to be, when in the natural state of their existence. I would refer to this exploitation as "Anthropogenic Greed". There is no better example, in the realm human history/memory, than the ruthless exploitation of land/soil resources, through the highly soil extractive, chemically-centric farming, euphemistically called the "green revolution", that left more than a third of India's geographical land area with degraded soils, in about two decades time, when the green revolution held it's sway in India.

When an ecosystem evolves the natural way, at a certain point, of its evolutionary cycle, it attains its climax. All natural ecosystems on planet Earth, which are virgin ones, are in the state of climax of their evolution. When in its climax, an ecosystem regulates itself through self-regulating mechanisms, then, it is at the highest level of function, contributing enormously to building up a life—ameliorating microclimate. On the other hand, anthropogenic ecosystems (such as agricultural systems)—when the human intervention is far too intensive—gradually advance towards its disclimax. At its disclimax, an ecosystem is not as effective/functional, as it ought to be, as at its climax.

When an ecosystem has attained its climax, it thrives with maximum photosynthetic efficiency and, reverberates with fullness of life. Intrinsic ecological factors operating within an ecosystem in its climax, provide bright chances for new species to evolve over an ecological time scale. Every new species of a living organism has evolved in such climax environments, adding the very richness to life. In an anthropogenic ecosystem, depending on the extent of human intervention, photosynthetic efficiency would be substantially reduced and paucity and crisis of life forms will prevail. In such an ecosystem, the chances of extinction of species is, indeed, very high.

In an ecosystem, where the photosynthetic efficiency is very high, it would result in higher primary production. The higher the rate of primary production, the more efficient would be the ecosystem function. The higher efficient the ecosystem function is, the higher the productivity of the specific ecosystem. Hence, the natural or climax ecosystem would reach maximum primary productivity, and the anthropogenic ones, unlike those of the former case, will perform, rather poorly. Sometimes, even an anthropogenic ecosystem could demonstrate higher productivity, provided, its native species are altered by monocultures of high yielding species, supplemented liberally, by external inputs, such as water, fertilizers, pesticides and herbicides. However, such transformed/artificially altered ecosystems are extremely vulnerable, during the passage of time, hence, their productivity is unsustainable, over a lengthy period of time. This is well demonstrated by the failure of the green revolution, after an initial spurt, when the ecosystem became vulnerable to pest and disease attack, consequently, the crop yield plateauing or declining.

What Are the Biodiversity-Based Sustainability Principles?

Biodiversity should not be mistaken for a biophysical aspect of life on planet Earth. Equally, it is, not the "Spice of Life", as some people would like to call it. As an outcome of natural evolution, biodiversity is a phenomenon, in itself, as well as the crux of life. The processes of life's sustenance continue via biodiversity. The sustainability of life is nurtured, basically, through Nature's biodiversity. Without the biodiversity, there never can be any sustainability, as the former is the central component. The following are the three vital principles of sustainability, all of which, revolve around the central component, biodiversity:

1. Soil Biodiversity
2. Above-soil biodiversity
3. Nutrient Cycles.

What is Soil Biodiversity?

Most, if not all, when a discussion takes place with regard to biodiversity, the vision is just confined to plants, and, it does not go beyond it. Soil should never be mistaken, merely, as a substratum for the plants to anchor their roots. It is much beyond that. Soil is a living ecosystem, and this has been adequately and scientifically, discussed in the book of this author titled "The Living Soil" (Nair, 2023a). Biodiversity of soil or the pedobiodiversity (pedo-flora and pedo-fauna)—is the basis of a living soil. There is a far greater measure of biodiversity per unit area and volume of the soil, below the surface, than that above the soil surface, visible in the troposphere. The biodiversity of the soil is fed by both chemosynthesis and photosynthesis. It is the living soil from where the structure of the biosphere, through the structure of all living beings, comes into being. This occurs through atmospheric nitrogen fixation, through the symbiotic bacteria, and, nitrogen is the fundamental constituent of all proteins, via the building blocks of the proteins, which are the amino acids. The phenomenon is attributable to the role of the nitrogen-fixing microorganism, the *Rhizobium* species in the soil. This microorganism occurs freely in the soil, but, with a specific association of only legume plants. The nitrogen is absorbed and assimilated by plants in the form of nitrates.

Soil biodiversity, from the viewpoint of its role in biological nitrogen fixation, would suggest that it is of a great and vital significance to the above-soil surface biodiversity. In other words, it would imply that the above—soil surface biodiversity is equally well nurtured by the below-soil surface biodiversity. The soil biodiversity, that apart from the N-fixing mechanism, includes various types of decomposers (microorganisms), which also influence the dynamics of nutrient cycles in the lithosphere, which is connected to all environmental components. Availability of all plant nutrients, major, minor and trace, in the soil determines the nutrition status, and, therefore, the health and well-being of the above-soil biodiversity. In conclusion, it can be said that the status of above-soil biodiversity (in the troposphere) is directly proportional to the below-soil biodiversity. From a larger perspective, it can be said that the well-being of all life, above the soil surface, including mankind, is closely linked to the well-being of the biodiversity below the soil surface.

The Biodiversity on the Surface of the Soil

For various socio-economic needs we always lay emphasis on the above soil surface biodiversity, namely, what one observes in the troposphere. This biodiversity is the basis of all of mankind's management practices—conservation and utilization. The higher the biodiversity above the soil/land surface, the greater the degree of sustainability—for all life, mankind, plant and animal. Photosynthesis is the basis for the survival of this biodiversity. All terrestrial consumers, ranging from herbivores to top of the rung carnivores (which includes mankind), are dependent on this biodiversity. In fact, humanity, directly depends on this biodiversity.

Soil biodiversity is nourished by the above the soil biodiversity. This takes place, principally, through the supply of organic matter to the soil. Organic matter content in the soil is nearly two to three fold more than that found in the vegetation outside the soil, but, most of it comes through the photosynthetic process by green plants. A small fraction of carbon is contributed through the process of chemosynthesis by soil microorganisms and photosynthesis by algae and blue green algae. Soil organic matter supports a huge number of high level diversity of detritivores. There is a close and deep interconnection between chemosynthetic and photosynthetic organisms, consumers, and detritivores, proving that all of the soil inhabiting organisms thrive through a healthy inter-relationship.

The Nutrient Cycles

Mineralization is the process by which nutrients contained in the organic matter are released into the soil matrix through action of microorganisms. While nitrogen and carbon get lost in their atmospheric pool, other nutrients, both metallic and non metallic, macro as well as micro nutrients, stay in their soil pool. These nutrients are made available to the plant growing above the soil. It is through thermodynamic action that these nutrients are made "bio available" to the plant roots. The thermodynamics of nutrient bioavailability is thoroughly discussed by in the book "Intelligent Soil Management for Sustainable Agriculture—The Nutrient Buffer Power Concept" authored by Nair (2019a). A portion of the nutrients also enrich the soil. The nutrient cycles, both sedimentary and gaseous, in this manner, nourish biodiversity within and out the soil. The nutrient cycles, which on the broader geological and geographical scales, on a global scale, which are referred to as bio-geochemical cycles, are vital for the very ecological integrity of planet Earth, and the Planet's biodiversity plays a key role in maintaining these cycles. The three biodiversity—centric sustainability principles, and how they are linked with various ecosystem functions is depicted in the following figure.

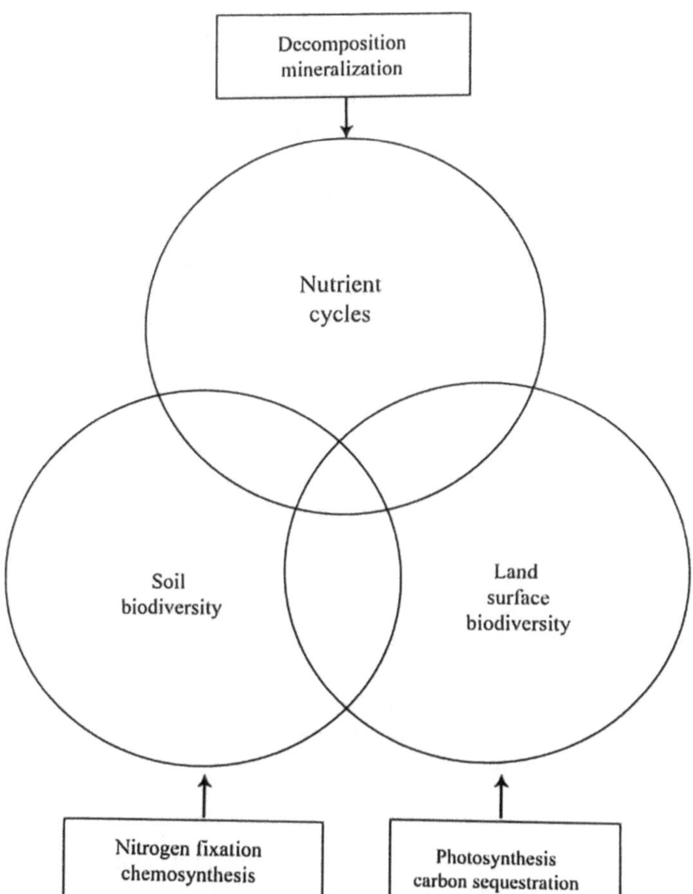

The three interlinked biodiversity-centric sustainability principles. Biodiversity as the main ecosystem structural component performs key ecosystem functions, namely, nitrogen fixation, photosynthesis, chemosynthesis, carbon sequestration, decomposition/mineralization of organic matter

How Do We Manage Sustainability?

When the ecosystems are natural, they attain ecological climax during their journey of evolution. When anthropogenic interference adversely affects the natural evolution, sustainability cannot be attained. We have quite a number of such instances in human history. Ecological regeneration has to precede degradation to make sustainability possible. The need for sustainability management arises in the case of human-managed systems, which are the ecosystems transformed to implement the various socio-economic activities of the society/community. If the first and foremost principle of sustainability, which is ecological sustainability, is unmet, we cannot attain a sustainable socio-economic system, in the case of agriculture, animal husbandry, gardening, industry, mining etc.

Within a socio-economic system, the ecological dimension of sustainability must be as big as it could be, to be able to compensate for the damage done to ecological parameters, for specific purposes, such as, a construction activity averting ecological regeneration of a land area. In the case of agriculture, a forest ecosystem serves as a core for establishing a sustainable ecosystem. Agriculture based on cultivated land and cut off from forests cannot be sustainable. Forest as an integral component of an agroecosystem or a farming system serves as a vibrant and healthy source of nutrients for the cultivated land.

There are many flower pollinating insects in a forest ecosystem, which through their role on pollination, help maintain the ecological integrity of the ecosystem. A forest is ecologically more stable than a cultivated field, where planted food crops with shallow roots have to depend only on the top soil for their nutrition. In other words, cultivated lands are more fragile than forests, inasmuch as sustainability management/or maintenance of the ecosystem is concerned. A natural forest with deep-rooted trees as the dominant biotic component, naturally, embraces a higher degree of biodiversity than what happens in a cultivated field, where the biodiversity (agrobiodiversity) is quite narrow. Biodiversity is the basis of ecological sustainability. In a forest-cultivated land, links in agriculture, thus, contribute to enhancing and maintaining sustainability of the entire ecosystem.

Rains conduct nutrient flow in a forest land. Also, animals, especially the birds, which feed on forest products, such as fruits, flowers and insects etc., act as nutrient conductors. In many developing economies like India, farmers manage such nutrient flow from forests to cultivated lands through livestock. Livestock feed on forest products, such as tree leaves, and the nutrients contained in them flow into the cultivated field when the farmers use the animal dung/manure to fertilize the crops sown in the field. Thus, livestock help maintain the flow of the nutrient cycle. They consume crop residues, especially straw and green fodder, and, a portion of consumed nutrients, after the animals utilize the bulk of the nutrients contained in straw and green fodder for their metabolic purposes, emanating from the straw/green fodder, get recycled into cultivated land through animal manure. Households, which are the major consumers of the cultivated produce, as well as forest products, also generate a lot of biodegradable/recyclable wastes. Such wastes contain a bulk of essential plant nutrients. After processing of these biodegradable/recyclable wastes, into composts or vermicompost, nutrient recycling from one source to another takes place, when these products, namely, composts/vermicomposts, are used to fertilize crop fields. Thus, nutrient cycling takes place in the ecosystem, which, in the long run, help maintain ecological sustainability.

What Are the Salient Aspects of Human Management Aimed to Sustain Biodiversity?

1. The Nature's biodiversity is enhanced through forests, hence, protect them and conserve them
2. In cultivated fields, agrobiodiversity is enhanced through the following measures:
   (a) Mixed cropping and multiple cropping
   (b) Growing deep-rooted crops combined with shallow-rooted ones
   (c) Increase genetic diversity of each crop
   (d) Through the adoption of agroforestry.
3. Adoption of animal husbandry which involves multiple livestock species and a variety of breeds of each livestock species
4. Through effective use of waste management, as for example, converting household wastes and animal manure into compost/vermicompost and applying the same onto cultivated field crops which would help enhance soil fertility and soil biodiversity.

For an ecologically sound, environmentally safe, functionally vibrant, and productive agricultural system, one must envision the establishment of an agroecosystem which involves natural forests, cultivated land, livestock and households in consonance with each other operating on the basis of sustainability principles. The natural forest area must, at least be, ten times bigger than the cultivated land area. The natural forest would help maintain soil, water, and biodiversity conservation, moisture circulation, carbon sequestration, and help maintain good and stable microclimate of an area, which, overall, ensures a conducive environment for the efficient functioning of an agro-ecosystem.

References

Anubhav, K. S., Rastogi, A., & Singh, V. (2009). Sacred groves in India: Celebrating sanctity of life through biodiversity conservation. In P. L. Gautam, V. Singh, & U. Melkania (Eds.), *Ecosystem diversity and carbon sequestration: Climate change challenges and a way out for ushering in a sustainable future* (pp. 89–101). Daya Publishing House.

IUCN-UNEP-WWF. (1980). *World conservation strategy: Living resource conservation for sustainable development* (77 pp). IUCN.

IUCN. (2019). *Guidelines for using the IUCN red list categories and criteria: Version 14* (113 pp). IUCN.

Mittermeir, R. A., Myers, N., & Gil, P. R., & Mittermeir, C. G. (1999). *Hotspots: Earth's biologically richest and most endangered terrestrial ecoregions.* Cemex, Conservation International and Agrupacion Sierra Madre.

Myers, N. (1988). Threatened biotas: "Hot spots" in tropical forests. *The Environmentalist, 8*, 187–208. https://doi.org/10.1007/BF02240252

Myers, N. (1990). The biodiversity challenge: Expanded hotspots analysis. *Environmentalist, 8*, 187–208. https://doi.org/10.1007/BFO2240252.

Myers, N., Mittermeir, R. A., Mittermeir, C. G., da Fonseca, G. A. B., & Kent, J. (2000). Biodiversity hotspots for conservation priorities. *Nature, 403*, 853–858. https://doi.org/10.1038/35002501

Nair, K. P. P. (2019a). *Intelligent soil management for sustainable agriculture—The nutrient buffer power concept.* Springer Nature Switzerland AG.

Nair, K. P. P. (2019b). Utilizing crop wild relatives to combat global warming. *Advances in Agronomy, 153*, 175–258.

Nair, K. P. P. (2019c). *Combating global warming—The role of crop wild relatives for food security.* Springer Nature Switzerland AG.

Nair, K. P. P. (2023a). *The living soil: A lifetime journey in understanding it for human sustenance. SpringerBriefs in Environmental Science.* Springer Nature Switzerland AG.

Nair, K. P. P. (2023b). *Extractive farming or Bio farming? Making a better choice for the 21st century. SpringerBriefs in Environmental Science.* Springer Nature Switzerland AG.

Ormsby, A. A., & Bhagavat, S. A. (2010). Sacred forests of India: A strong tradition of community-based natural resource management. *Environmental Conservation, 37*(3), 320–326. https://doi.org/10.1017/S0376892910000651

Singh, V. (2019). *Fertilizing the universe: A new chapter of unfolding evolution* (285 pp). Cambridge Scholars Publishing.

Singh, V. (2020). *Environmental plant physiology: Botanical strategies for a climate smart planet* (230 pp). Taylor and Francis (CRC Press).

MIX
Papier aus verantwortungsvollen Quellen
Paper from responsible sources
FSC® C105338

If you have any concerns about our products,
you can contact us on
ProductSafety@springernature.com

In case Publisher is established outside the EU,
the EU authorized representative is:
Springer Nature Customer Service Center GmbH
Europaplatz 3, 69115 Heidelberg, Germany

Printed by Libri Plureos GmbH
in Hamburg, Germany